AA007699

11th International Conference on Mechatronics and Manufacturing (ICMM 2020)

IOP Conference Series: Materials Science and Engineering
Volume 895

Tokyo, Japan
12 January 2020

ISBN: 978-1-7138-2692-7
ISSN: 1757-8981

Printed from e-media with permission by:

Curran Associates, Inc.
57 Morehouse Lane
Red Hook, NY 12571

Some format issues inherent in the e-media version may also appear in this print version.

Institute of Physics
Dirac House, Temple Back
Bristol BS1 6BE UK

Phone: 44 1 17 929 7481
Fax: 44 1 17 920 0979

techtracking@iop.org

Additional copies of this publication are available from:

Curran Associates, Inc.
57 Morehouse Lane
Red Hook, NY 12571 USA
Phone: 845-758-0400
Fax: 845-758-2633
Email: curran@proceedings.com
Web: www.proceedings.com

TABLE OF CONTENTS

MATERIAL STRUCTURE DESIGN AND PERFORMANCE ANALYSIS

MECHANICAL AND ELECTRONIC CONTROL ENGINEERING

IOP Conf. Series: Materials Science and Engineering **895** (2020) 011001 doi:10.1088/1757-899X/895/1/011001

Preface

Welcome to the 2020 11th International Conference on Mechatronics and Manufacturing (ICMM 2020), which was held at Tama Campus, Chuo University, Tokyo, Japan during January 10-12, 2020.

The ICMM 2020 had provided a forum for scientific and technological advances in the theory and practice of mechatronics and manufacturing. In recent years, the development of sensors and actuators promoted many new and challenging research fields of mechatronics, such and robotics, MEMS, IT and so on. Furthermore, new research trends are developing in the field of manufacturing, such as ultra-precision machining, additive manufacturing, multi-material systems and so on. Consequently, holding a highly-professional meeting, in which many academic researchers and industrial engineers could exchange their knowledge and experiences, is getting more and more important.

ICMM is growing with the numbers of participants and increasing in its quality. The latest ICMM 2020 aimed to be one of the leading international conferences in the world, and it provided an exciting environment for researchers and engineers to present and discuss the latest technologies and applications in the field of mechatronics and manufacturing.

In the present conference, a number of academic and industrial works have been submitted and selected after a peer review process by experts among the technical committee members and other international reviewers in the related fields of expertise. We hope all the attendants had experienced a remarkable opportunity for the academic and industrial communities to address new challenges and share solutions, and discuss future research directions.

Finally, we thank all participants who had come to Tama and joined the conference.

Conference Chair of ICMM 2020
Yoshihiko UEMATSU
Gifu University, JAPAN

IOP Conf. Series: Materials Science and Engineering **895** (2020) 011002 doi:10.1088/1757-899X/895/1/011002

Conference Committee

Advisory Committee

Prof. Makoto IWASAKI, Nagoya Institute of Technology, Japan

Conference Chairs

Prof. Yoshihiko Uematsu, Gifu University, Japan
Assoc. Prof. Ratchatin Chancharoen, Chulalongkorn University, Thailand

Program Chairs

Prof. Mahir Dursun, Gazi University, Turkey
Prof. Vadim R. Gasiyarov, South Ural State University, Russia

Technical Committee

Prof.Shailendra Kumar, S.V.National Institute of Technology, India

Prof.Bandit Suksawat, King Mongkut's University of Technology North Bangkok, Thailand

Prof. Ranganath M. S, Delhi Technological University, India

Prof. Prakash.S, Sathyabama University, India

Prof. Osman Adiguzel, Firat University, Turkey

Prof. Ghassan A. Al-Kindi, Sohar University - Sultanate of Oman

Prof. Anna A. Sandulyak, Moscow technological university, Russia

Prof. N.Ethiraj, Dr. M.G.R Educational and Research Institute, India

Prof. Hamid Reza Karimi, Polytechnic University of Milan, Italy

Assoc.Prof. Tapas Kumar Maiti, Hiroshima University, Japan

Assoc.Prof. K Kanlayasiri, King Mongkut's Institute of Technology Ladkrabang, Thailand

Assoc. Prof. Seung-Jun Shin, Hanyang University, South Korea

Assoc. Prof. Dr. V. Anandakrishnan, National Institute of Technology, India

IOP Conf. Series: Materials Science and Engineering **895** (2020) 011002 doi:10.1088/1757-899X/895/1/011002

First name	Surname	Institution	E-mail address
Adirek	Baisukhan	Chiang Mai University	adirek_b@cmu.ac.th
A	Godfrey	Tsinghua University, China	awgodfrey@mail.tsinghua.edu.cn
Sai Kiran Chary	Nalband	Mahatma Gandhi Institute of Technology, India	charysaikiran1@gmail.com
Chunhui	Dai	Wuhan Second Ship Design and Research Institute, China	ch.dai@163.com
Xingsheng	Lao	Science and Technology on Thermal Energy and Power Laboratory, Wuhan 2nd Ship Design and Research Institute, China	460079620@qq.com
Beitong	Zhou	Huazhong University of Science and Technology, China	zhoubt@hust.edu.cn
Dhiren R.	Patel	Department of Mechanical Engineering, School of Technology, PDPU, India	dhirenpatel85@gmail.com
Sunti	Thaitanun	Chulalongkorn University, Thailand	gridsada.phanomchoeng@gmail.com
Phuoc Vinh	Dang	The University of Danang, Danang University of Science and Technology, Vietnam	dpvinh@dut.udn.vn
Jirawath	Parnklang	King Mongkut's Institute of Technology Ladkrabang Bangkok, Thailand	jirawath.par@kmitl.ac.th
Boris	Gurenko	Southern Federal University, Rostov-on-Don, Russia	boris.gurenko@gmail.com
Mucahid	Barstugan	Konya Technical University, Turkey	mbarstugan@ktun.edu.tr
Apinai	Rerkratn	King Mongkut's University of Technology North Bangkok, Thailand	apinai@gmail.com
Warisa	Wisittipanich	Chiang Mai University, China	warisa.o@gmail.com
Sutham	Satthamsakul	King Mongkut's Institute of Technology Ladkrabang, Bangkok, Thailand	sutham.sa@kmitl.ac.th
M P	Suriyage	University of Moratuwa, Sri Lanka	mpsuriyage@gmail.com

IOP Conf. Series: Materials Science and Engineering **895** (2020) 011004 doi:10.1088/1757-899X/895/1/011004

Peer review statement

All papers published in this volume of *IOP Conference Series: Materials Science and Engineering* have been peer reviewed through processes administered by the proceedings Editors. Reviews were conducted by expert referees to the professional and scientific standards expected of a proceedings journal published by IOP Publishing.

IOP Conf. Series: Materials Science and Engineering **895** (2020) 012001 doi:10.1088/1757-899X/895/1/012001

Effects of Friction Stir Welding Parameters of Dissimilar Aluminum Alloys on Residual Stress and Microhardness

Adirek Baisukhan[1,a], **Wasawat Nakkiew**[1,b*]

[1]Advanced Manufacturing Technology Research Center (AMTech),
Department of Industrial Engineering, Faculty of Engineering,
Chiang Mai University, Thailand

[a]adirek_b@cmu.ac.th, [b]wasawat@eng.cmu.ac.th ([*] Corresponding author)

Abstract. The aim of this research is to use a full factorial design to determine the significant factors of friction stir welding (FSW) process of dissimilar AA6061-T6 and AA7075-T651 aluminum alloys, on the surface residual stress and microhardness. There are a total of three factors studied: rotation speed, welding speed and workpiece layout. The results showed that the factors that have a statistically significant effect on residual stress occurring are welding speed, workpiece layout and interaction between rotation speed and welding speed. The appropriate parameters for FSW process of dissimilar aluminum alloys are rotation speed of 1400 rpm, welding speed of 50 mm/min and the layout of the workpiece by advancing side (AS) uses AA6061-T6 sheet, while retreating side (RS) uses AA7075-T651 sheet, which will result in the surface residual stress of -34.33 MPa. This compressive residual stress will be beneficial to the welded joint for retard the occurrence of cracks caused by fatigue. The hardness of all workpieces have similar tendency, thermo-mechanically affected zone (TMAZ) area in both AS and RS gave the least hardness, which found that the AA6061-T6 and AA7075-T651 sheets had a hardness of approximately 55HV and 110HV respectively.

1. Introduction

Friction stir welding (FSW) has been invented and developed by The Welding Institute (TWI), United Kingdom, 1991. FSW is widely used in automotive, shipbuilding, trains and aerospace applications, which mainly used in the welding of aluminum alloys and can be welded to dissimilar materials. On the other hand, it is well known that welding different types of aluminum alloys is difficult to weld using the traditional fusion welding [1, 2]. Therefore, the welding of different types of aluminum alloys is preferable to use FSW, which consists of many factors such as rotation speed, welding speed, characteristics of welding tool, tilt angle, etc. These factors all affect the properties and quality of the welded joint. Now a day, these factors are still being studied continuously [3–5]. One of the most interesting response of welded joint is the surface residual stress as well as microhardness, as they play an important role in cyclic applications where most of the welds are damaged due to fatigue [6, 7]. The compressive residual stress will be beneficial, which increases fatigue life and retard cracks initiation. Thus, residual stress determination is performed using X-ray diffraction (XRD) method, which is widely used method and does not need to destroy the workpiece. The experimental design method is a powerful method that can be used to study the factors that affect surface residual stress and microhardness [8–10]. In this research, the 2^3 full factorial design experiment will be used for determining significant factors of three FSW process parameters: rotation speed, welding speed and workpiece layout, to the residual stresses occurred and microhardness after the FSW process.

2. Experimental Procedure

2.1. Materials

This research uses dissimilar materials which are AA7075-T651 and AA6061-T6 aluminum alloy, with each piece having a width of 100 mm, length of 200 mm and thickness of 6.5 mm. The chemical composition measured by the energy dispersive X-ray fluorescence (EDXRF) method, as shown in Table 1. The prepared workpiece will be welded in a single pass and using a square butt joint. For FSW tools made from tool steel AISI H13 or equivalent to DIN1.2344 or JIS SKD61, which have 6x6 mm square pins, the length of pin 6 mm and the tool shoulder 18 mm, this tool has a hardness of approximately 55 HRC after the heat treatment process.

Table 1. Chemical composition of AA7075-T651 and AA6061-T6 aluminum alloys.

Alloys	Elements (% w/w)							
	Cu	**Mn**	**Mg**	**Zn**	**Cr**	**Fe**	**Si**	**Al**
AA7075-T651	1.83	0.09	4.95	7.39	0.32	0.37	-	Balance
AA6061-T6	0.34	0.16	-	0.15	-	0.64	0.84	Balance

2.2. Design of Experiment of Friction Stir Welding

In this research, the FSW process was done with the CNC machining center Bridgeport VMC500 and the schematic diagram of FSW process, as shown in Figure 1. For the design of this experiment used a full factorial design of 2^3 by 2 replication total of 16 experiments with three factors: rotation speed, welding speed and workpiece layout each factor consists of a low and high levels where the desired response is the surface residual stress at the middle of the weld beat, as shown in Table 2.

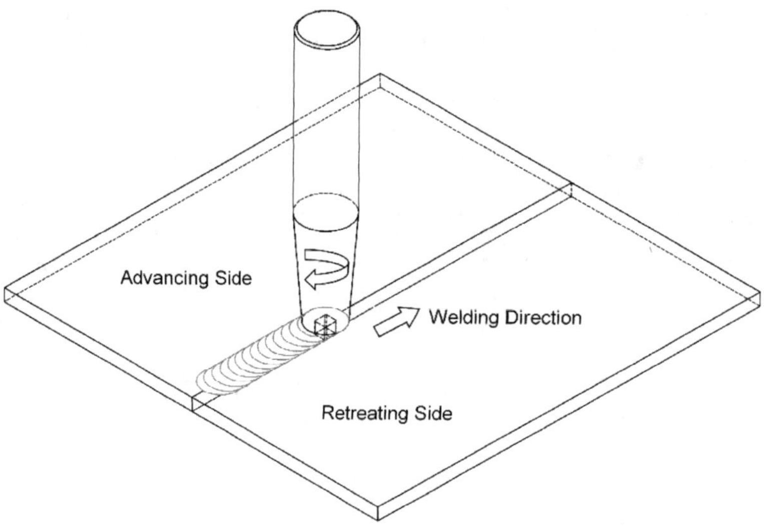

Figure 1. Schematic diagram of FSW process.

Table 2. Full factorial design of FSW process.

Factors	Unit	Level	
		Low (-1)	**High (+1)**
Rotation speed	rpm	1,000	1,400
Welding speed	mm/min	30	50
Workpiece layout	-	Layout A[a]	Layout B[b]

[a] AS uses a piece AA6061-T6 and RS uses a piece AA7075-T651
[b] AS uses a piece AA7075-T651 and RS uses a piece AA6061-T6

2.3. Residual Stress Measurement

For surface residual stress from FSW workpieces will be obtained from XRD with $\sin^2\psi$ method, which is a non-destructive method, by Stress-tech XSTRESS 3000 equipped with a Chromium tube source radiation. Eleven tilt angles were used both positive and negative values (0, 18.4°, 26.6°, 33.2°, 39.2°, 45°) and temperature during measurement about 25 °C. The location for measurement surface residual stresses by using XRD is at the welding center line of longitudinal direction.

2.4. Microhardness Measurement

Vickers microhardness value of workpieces will be obtained from Startec SMV-1000 machine with apply force about 1.961N or HV0.2, holding time 10 sec and using a magnification of 100X for measuring the width of the pressing mark. Then, preparing and polishing the cross section area of workpiece for measure the hardness from the center of the friction stir welding to the left and right, as shown in Figure 2. Each side measured 6 positions, totalling 13 positions, each position 2 mm apart and measured two times to find the average.

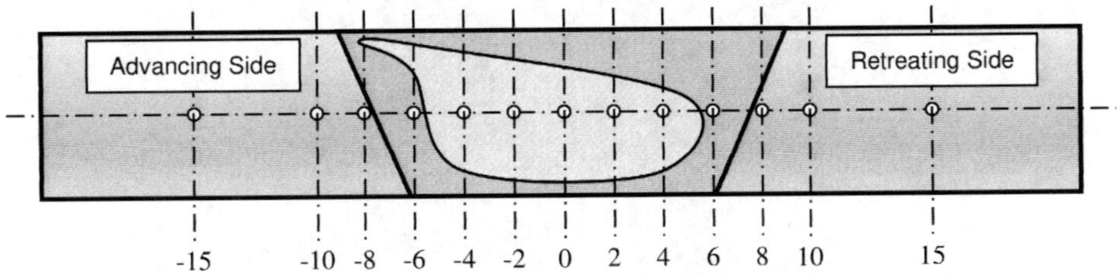

Figure 2. Position for measuring microhardness.

3. Results and Discussion

3.1. Residual Stress and Experimental Design Results

From a visual inspection through the microscope, it was found that the welded area of all FSW workpiece according to the experimental design had no defects occur, both aluminum alloys are well bonded, no cracks or bubbles are found, as shown in Figure 3. For the result of surface residual stress obtained from the XRD measurements according to the full factorial design measuring were found to have both tensile residual stress (positive value) and compressive residual stress (negative value), as shown in Table 3, which is different from the traditional fusion welding (MIG or TIG welding) that the middle of the weld beat gives tensile residual stress. This FSW workpieces shows that the welded joint is likely to be a longer fatigue life than the welded joint by traditional fusion welding methods due to surface residual stress. The analysis of experimental design was performed by the statistical software MINITAB, according to the results, as shown in Figure 4, indicating that welding speed, workpiece layout and interaction between rotation speed and welding speed were statistically significant factors (p-value < 0.05), the adjusted coefficient of determination (R^2) was 68.47%, indicating that the model was sufficient to fit the data. The validation of the data, as shown in Figure 5. The Normal Probability Plot graph found that the residual data of the residual stress tends to be linear because the points of the residues are arranged in a straight line. The Histogram graph found that the residual data of the residual stress values had a normal distribution, which the characteristics of the graph were similar to the bell-shaped. The Versus Fits graph found that the residual data were scattered evenly and do not have a clear trend pattern, therefore the residual data have independent of predicted values. The Versus Order graph found that the residual data were distributed consistently and did not have a clear trend, therefore the residual data have independent of the experiment sequence. The results of the analysis of variance (ANOVA), as shown in Table 4, found that the lack of fit was insignificant because the p-value of lack of fit is greater than the significance level 0.05, meaning that the equation is suitable for the experimental results and for the p-value (at the significant level 0.05) of the main effects: welding speed, workpiece layout and 2-way interactions between rotation speed and welding speed were less than 0.05, therefore those effects were significant factors which then be used to construct the regression equation for predicting the surface residual stress. The equation for predicting the surface residual stress at the middle of weld beat obtained from the regression analysis of factors that have a significant, as shown in Equation (1).

$$\text{Residual stress} = -2.74 - 8.55(\text{Welding speed}) + 11.68(\text{Workpiece layout}) - 11.36(\text{Rotation speed} * \text{Welding speed}) \quad (1)$$

Using Equation (1), the surface residual stress was calculated by substitute the value +1, +1 and -1 to rotation speed, welding speed and workpiece layout respectively, which mean rotation speed = 1,400 rpm, welding speed = 50 mm/min and workpiece layout = layout A, then the most compressive surface residual stress value approximately -34.33 MPa

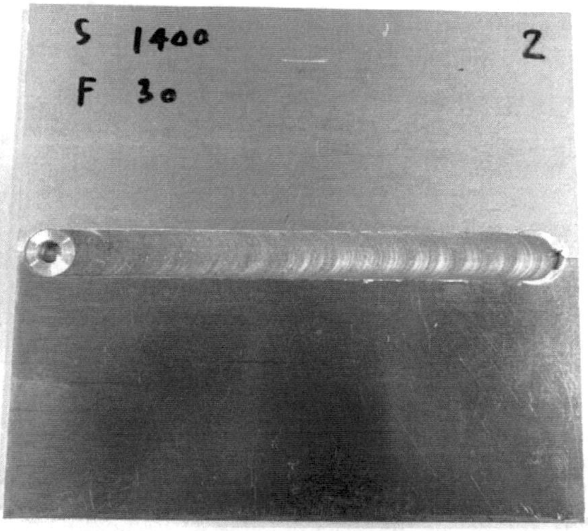

Figure 3. FSW workpieces.

Table 3. Result of surface residual stress.

Standard Order	Run Order	Rotation speed	Welding speed	Workpiece layout	Residual stress (MPa)
7	1	1,000	50	Layout B	5.3
14	2	1,400	30	Layout B	10.6
8	3	1,400	50	Layout B	-3.0
6	4	1,400	30	Layout B	32.4
2	5	1,400	30	Layout A	9.5
3	6	1,000	50	Layout A	-11.7
11	7	1,000	50	Layout A	-27.5
15	8	1,000	50	Layout B	19.5
9	9	1,000	30	Layout A	-14.8
4	10	1,400	50	Layout A	-21.0
5	11	1,000	30	Layout B	1.6
12	12	1,400	50	Layout A	-48.0
16	13	1,400	50	Layout B	-3.9
1	14	1,000	30	Layout A	-32.7
13	15	1,000	30	Layout B	9.0
10	16	1,400	30	Layout A	30.9

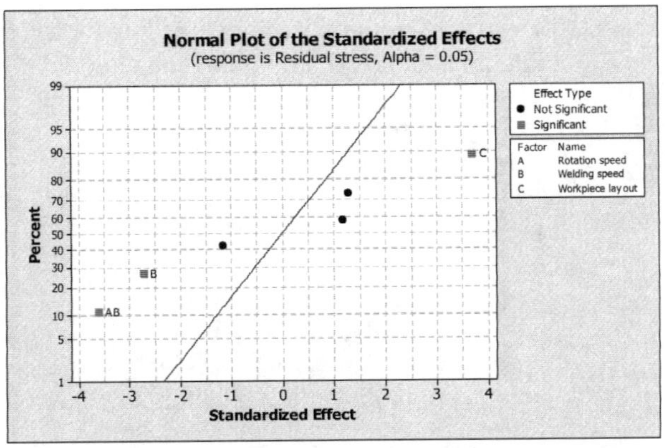

Figure 4. Normal plot of the standardized effects.

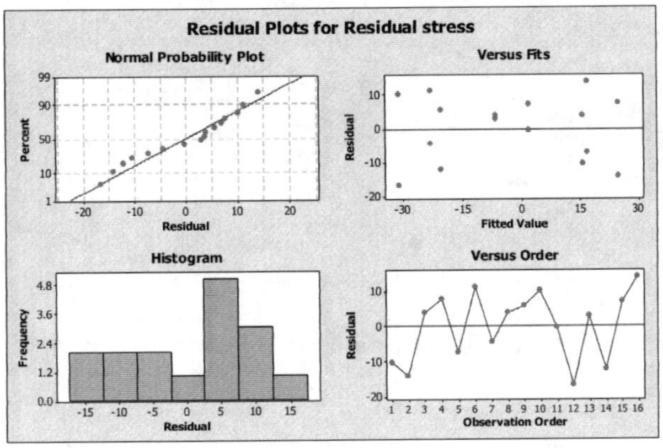

Figure 5. Residual plots for residual stress.

Table 4. Analysis of variance for residual stress (coded units).

Source	DF	Seq SS	Adj SS	Adj MS	F	P
Main Effects	3	3566.6	3566.6	1188.9	7.51	0.008
Rotation speed	1	216.1	216.1	216.1	1.37	0.273
Welding speed	1	1169.6	1169.6	1169.6	7.39	0.024
Workpiece layout	1	2180.9	2180.9	2180.9	13.78	0.005
2-Way Interactions	3	2538.9	2538.9	846.3	5.35	0.022
Rotation speed*Welding speed	1	2065.7	2065.7	2065.7	13.05	0.006
Rotation speed*Workpiece layout	1	205.9	205.9	205.9	1.3	0.283
Welding speed*Workpiece layout	1	267.3	267.3	267.3	1.69	0.226
Residual Error	9	1424.3	1424.3	158.3		
Lack of Fit	1	179.6	179.6	179.6	1.15	0.314
Pure Error	8	1244.7	1244.7	155.6		
Total	15	7529.9				

3.2. Microhardness Results

The base material (BM) of the workpiece, located at a distance of 15 mm, found that the workpiece AA6061-T6 has an average hardness of about 100HV that closed to a standard value, while the workpiece AA7075-T651 has an average value of about 150HV, which is slightly lower than the standard. The weld nugget area near the pin produces relatively consistent hardness with AS of the workpiece AA6061-T6 and AA7075-T651, the average hardness at this area is around 60HV and 130HV, respectively. On the RS of the workpiece AA6061-T6 and AA7075-T651 have hardness for this area about 60HV and 110HV respectively. The position with the least hardness, the range from 8 to 10 mm, which is the area of TMAZ on both the AS and RS of all workpieces is affected by the heat from the size of the welding tool shoulder, thus AA6061-T6 and AA7075-T651 will have an average hardness value around 55HV and 110HV respectively, as shown in Figure 6.

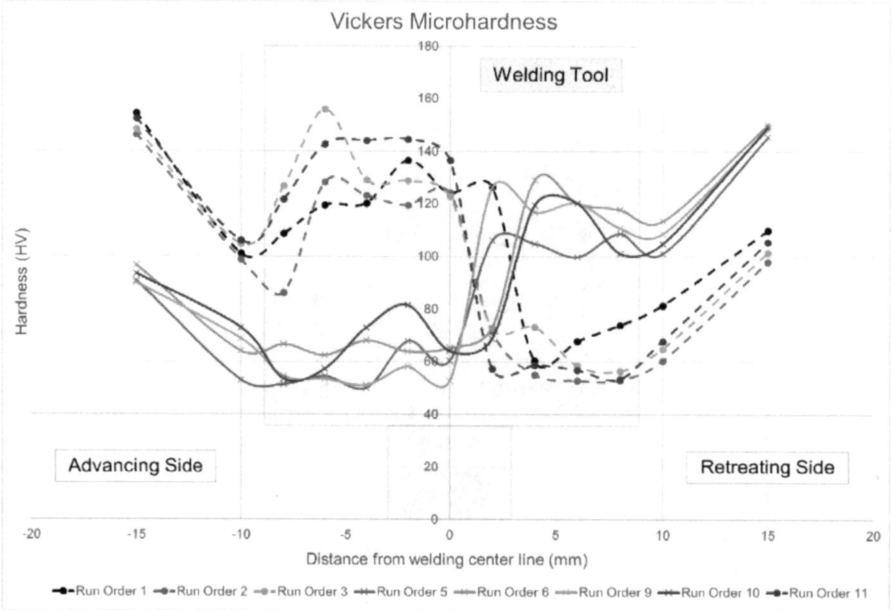

Figure 6. Results of Vickers microhardness.

4. Conclusions

The full factorial design was used for determining significant of FSW process parameters of dissimilar aluminum alloys to the occurrence of residual stresses and microhardness.

- Statistically significant factors of FSW process were welding speed, workpiece layout and interaction between rotation speed and welding speed.
- The appropriate FSW process parameters that will cause the minimum surface residual stress at the center of the weld beat approximately -34.33 MPa were rotation speed 1,400 rpm, welding speed 50 mm/min and workpiece layout that AS uses a piece of AA6061-T6 and RS uses a piece of AA7075-T651.
- The hardness of all workpieces have similar tendency, therefore, choosing the appropriate parameters must consider other properties together.

In the future, additional fatigue testing will be performed to find the S-N graph in order to compare stress values with the number of cycles of each workpiece.

5. References

[1] Hadji I, Badji R, Gaceb M, Kherrouba N and Rabahi L 2018 *Mater. Sci. Eng.* **461** 012022
[2] Muhsin J J, Kadhim K R, Esraa A A and Muhannad A 2018 *Mater. Sci. Eng.* **454** 012059
[3] Jamshidi H A 2015 *Mater. Des.* **87** 405–413
[4] Guo J F, Chen H C, Sun C N, Bi G, Sun Z and Wei J 2014 *Mater. Des.* **56** 185–192
[5] Rodriguez R I, Jordon J B, Allison P G, Rushing T and Garcia L 2015 *Mater. Des.* **83** 60–65
[6] Kadhim K R, Abbas A A, Muhannad A and Muhsin J J 2018 *Int. J. Mech. Eng.* **18** 112–120
[7] Juijerm P, Altenberger I and Scholtes B 2007 *Int. J. Fatigue* **29** 1374–1382
[8] Baisukhan A, Nakkiew W and Pitjamit S 2015 *Lect. Notes. Electr. En.* **349** 77–84
[9] Zapata J, Toro M and Lopez D 2016 *J. Mater. Process. Technol.* **229** 121–127
[10] Javadi Y, Sadeghi S and Najafabadi M A 2014 *Mater. Des.* **55** 27–34

Deformation behavior study in a model dual phase system of copper–martensitic steel using in-situ synchrotron X-ray diffraction

M Teimouri[1], A Godfrey[1*], B Yu[2] and N Tsuji[2]

[1] Key Laboratory of Advanced Materials (MOE), School of Materials Science and Engineering, Tsinghua University, Beijing 100084, China
[2] Department of Materials Science and Engineering, Kyoto University, Yoshida-honmachi, Sakyo-ku, Kyoto, 606-8501, Japan

awgodfrey@mail.tsinghua.edu.cn

Abstract. In this study, the deformation behavior of two phases in a model dual phase system of copper and a martensitic stainless steel was investigated using in situ synchrotron X-ray diffraction. Due to the different crystallographic structures of copper and martensite, their diffraction patterns are well separated such that the strain distribution and the load partitioning between the phases could be investigated under loading using standard methods. The copper matrix started to yield at low stress levels after a short stage of elastic deformation and the different lattice strain of the copper and martensite indicated that load partitioning took place after yielding of the copper.

1. Introduction

Dual-phase (DP) steels, typically composed of hard martensite particles dispersed in a soft and ductile ferrite matrix, show remarkable mechanical properties such as a combination of high strength and excellent formability. These properties make them attractive as engineering materials in the automobile industry and other structural applications. The mechanical behavior of DP steels arises from the partitioning of stress and strain between the two phases during deformation and depends on microstructural parameters such as grain size, distribution and volume fraction of the constituent phases. In order to understand the deformation characteristic in dual phase steels, an accurate evaluation of the micromechanical behavior of the constituent phases in such materials is required [1-4].

Synchrotron X-ray radiation can be applied to study the structural properties and deformation behavior of a wide range of engineering materials. As a powerful microstructural characterization tool, it can provide in situ analysis of the stress/strain partitioning between constituent phases and changes in lattice parameters during loading. Investigating the properties of the constituent phases is straightforward using such diffraction techniques if the two phases exhibit different crystallographic structures. However, as the diffraction patterns of the martensite and the ferrite/bainite overlap, owing to the similar crystal structures and lattice parameters of the two phases, analysis of the behavior of individual reflections is complicated and requires assumptions about the diffraction peak shapes and volume fractions of each phase [5-19].

In this article, we study the micromechanical behavior of a model dual phase steel prepared by sintering of mixed copper and AISI420 stainless steel powders and demonstrate that this material provides a good model system for investigating the stress and strain partitioning between constituent phases.

2. Experimental

Spherical powders of copper (with a purity of 99.9%) and AISI420 stainless steel with an average size of 3.5 and 10μm, respectively, were combined in weight ratio of 70:30 using a low-energy horizontal ball milling system at room temperature. The powder mixtures were milled for 24hrs. using stainless steel balls and a polymer vial at a speed of 120 rpm with a ball-to-powder weight ratio of 20:1.
Consolidation of the mixed powders was carried out using the spark plasma sintering (SPS) process. Sintering was performed under primary dynamic vacuum with an initial pressure of 7 MPa to maintain instrument stability. A sintering temperature of 1000°C was used in this study using, heated to 950°C at a rate of 100°C/min, followed by a ramp of 50°C/min for the final 50°C heating step. After reaching the final sintering temperature the pressure was increased to a load of 50MPa and held for 5 min. The samples were then unloaded and cooled under vacuum to room temperature. After sintering, the compacted samples were heat treated in order to control the martensitic transformation of AISI420 stainless steel phase. For this, the samples were heated to 1000°C at a heating rate of 10°C/min and held for 15min before quenching in oil. The samples were then tempered in an air-furnace at 500°C for 30 min.
In order to obtain diffraction measurements with suitable time resolution and without interruption of the deformation process, the synchrotron X-ray experiment was performed at a high-flux undulator beam line (BL46XU) at the SPring-8 synchrotron radiation facility. The X-ray energy was 30keV, corresponding to a wavelength of $\lambda = 0.413$ Å, and the incident beam was 0.5 and 0.2mm in the horizontal and vertical directions. The time resolution of this experiment was 1.0s. To allow simultaneous measurements over a wide range of diffraction angles with high angular resolution, an array of six MYTHEN detectors (a one-dimensional microstrip detector made by DECTRIS) was employed [12]. The tensile tests were performed with an initial strain rate of 8.3×10^{-4} s^{-1} at room temperature. Dog-bone shape specimens with a gauge length of 8 mm, width of 2 mm and thickness of 1 mm was used.
In order to obtain the parameters of interest such as integrated intensity, peak position and full width at half maximum (FWHM) values, the reflections of each recorded diffraction pattern were fitted using the Pseudo-Voigt function. The lattice strain distribution and evolution of different planes were investigated by analyzing the change of diffraction peak positions along 2D diffraction patterns of different phases. The lattice strain of the {hkl} planes, ε, was obtained using $\varepsilon = (d - d_0)/d_0$ where d and d_0 are the interplanar spacing of the {hkl} planes before and under loading, respectively, determined using Bragg's law, $\lambda = 2d\sin\theta$, where λ is the wavelength of the incident X-rays and θ is the diffraction peak angle. The dislocation density evolution of two phases during tensile loading was evaluated using the Williamson-Hall method:

$$\frac{\beta\cos\theta}{\lambda} = \frac{0.9}{D} + \frac{2\varepsilon\sin\theta}{\lambda}$$

where β and D are FWHM and the crystallite size, respectively. The values of D and ε were obtained from the slope and intercept of a plot of $\beta\cos\theta/\lambda$ against $\sin\theta/\lambda$ for each diffraction peak. The dislocation density, ρ, was then obtained according to $\rho = (k\varepsilon^2/\boldsymbol{b}^2)$ where ***b*** is the Burgers vector, k=16.1 for face centered cubic (FCC) materials, and k=14.4 for body-centered-cubic (BCC) metals [20-21].

3. Results and discussion

Structural characterization was performed with the aim to determine the copper and the martensite distribution homogeneity and morphology across the specimens using a scanning electron microscope (SEM). An example SEM micrograph is shown in figure 1, where the lighter continuous phase is the copper and the darker discrete regions are the AISI420 stainless steel. Good interfacial bonding between the two phases is seen, though with some porosity, also present in the copper phase.

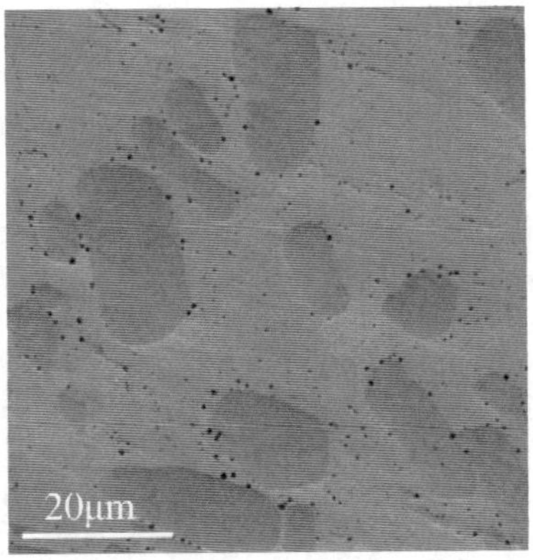

Figure 1. SEM image of the specimen.

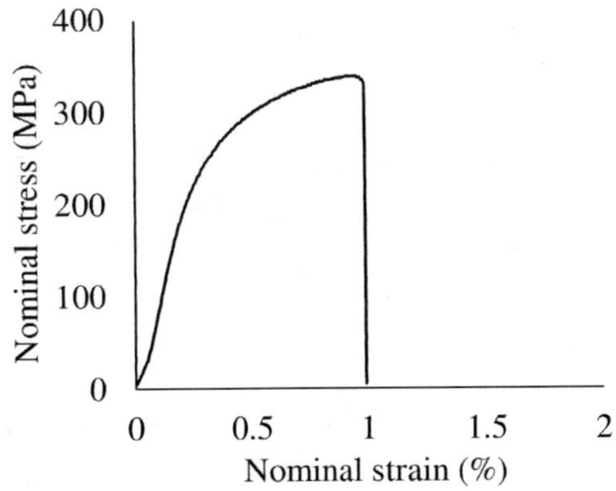

Figure 2. The engineering stress-strain curve of the sample.

Figure 2 shows the engineering stress-strain curve obtained during the tensile experiment. Each point represents a set of synchrotron measurements data were simultaneously acquired for the {111}, {200}, {220}, {311} and {222} peaks for copper, and the {110}, {200} and {211} peaks for the martensite phase. The specimen was strained up to 1% strain to ensure loading into a region of macroscopic plasticity.

Example one-dimensional XRD diffraction patterns of the as-received and under a load of 335 MPa deformed samples, with the measured direction parallel to the loading direction are shown in figure 3. Before loading, the specimen exhibits phases with two types of crystal structure: a BCC martensite phase and an FCC copper phase. With increasing deformation, the intensity of copper peaks decreases and weak new peaks appear, assumed here to be the ε phase.

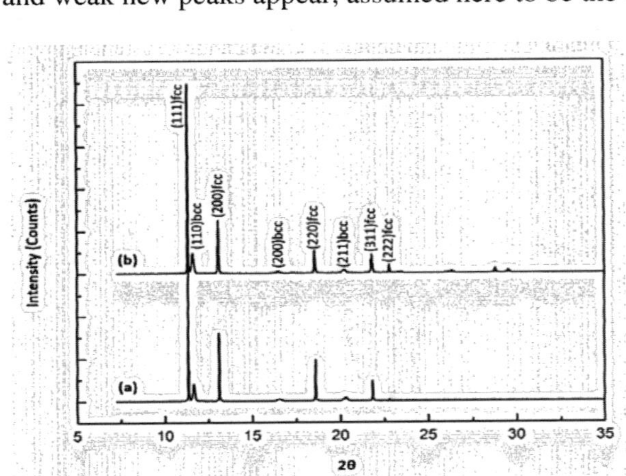

Figure 3. One- dimensional (1D) diffraction patterns of the model dual phase system at different applied stresses: (a) 0 MPa, and (b) 335 MPa.

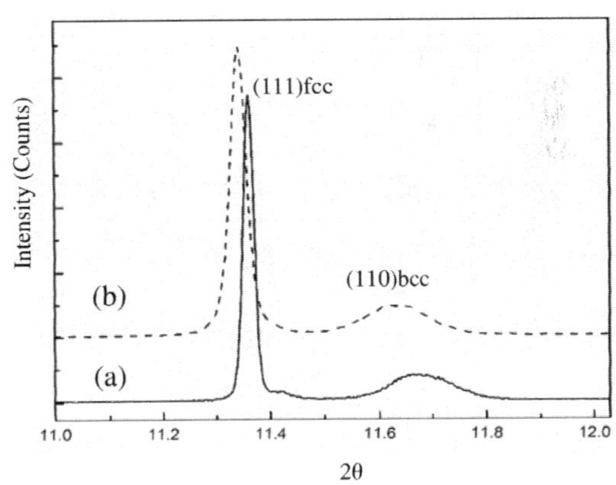

Figure 4. Diffraction peaks of copper and martensite at different applied stresses: (a) 0 MPa, and (b) 335 MPa.

Figure 4 shows the effect of strain on the {111} and {110} diffraction peak for copper and martensite, respectively. The shape of the diffraction peaks under loading is unchanged, but both peaks shift towards a smaller diffraction angle. This downward shift of the peak angle represents an increase in the plane spacing in the tensile direction as a result of elastic deformation. The martensite peak is more shifted, indicating a larger tensile stress in this phase.

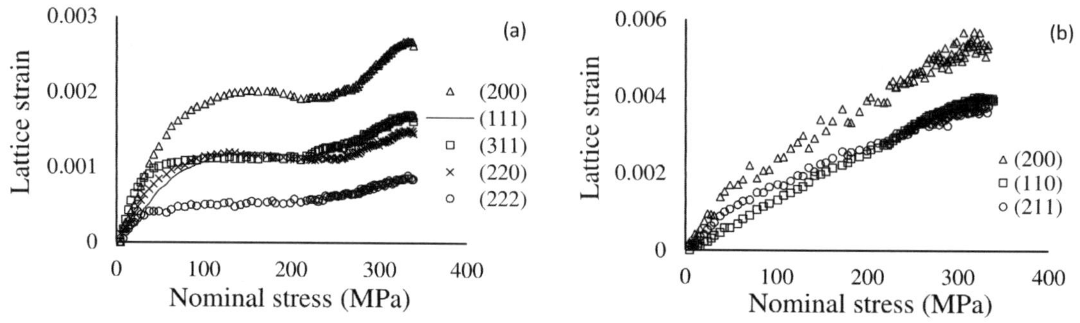

Figure 5. Changes in (a) copper and (b) martensite {hkl} lattice strains as a function of the nominal stress.

Figure 5(a,b) shows the measured lattice strains along the loading direction for different {hkl} planes of both the copper and martensite phases as a function of nominal stress. It can be seen that at the initial stage of loading, the lattice strains of the constituent phases increase linearly with applied stress, indicating that both phases deform elastically in this stage and different slopes are found for different reflections, indicating the different crystal orientations and anisotropy of the Young's modulus. When the stress reaches ~25 MPa, the lattice strain in the copper phase shows a non-linear dependence on the applied stress, while the hard martensite phase continues to deform elastically, indicating a change in the stress partitioning between the two phases. It can be seen that the lattice strain of the martensite planes continuously increases during almost the entire applied deformation and the martensite bears a much higher stress than copper, which always has a lower overall lattice strain value.
The behavior of lattice strain is dependent on crystal orientation, and yielding does not take place simultaneously in all the {hkl} lattice planes. In the copper phase, the {311}, {220} and {111} planes show a similar lattice strain at each stress, though the {311} plane shows evidence of yielding at a lower stress than the other two planes. The lattice strain of the {200} planes is highest, while the {222} planes show the lowest lattice strain, and also show evidence of earlier plastic yielding in the copper phase. For the martensite phase, at any given stress more load is carried by the {200} planes than by the {110} and {211} planes.

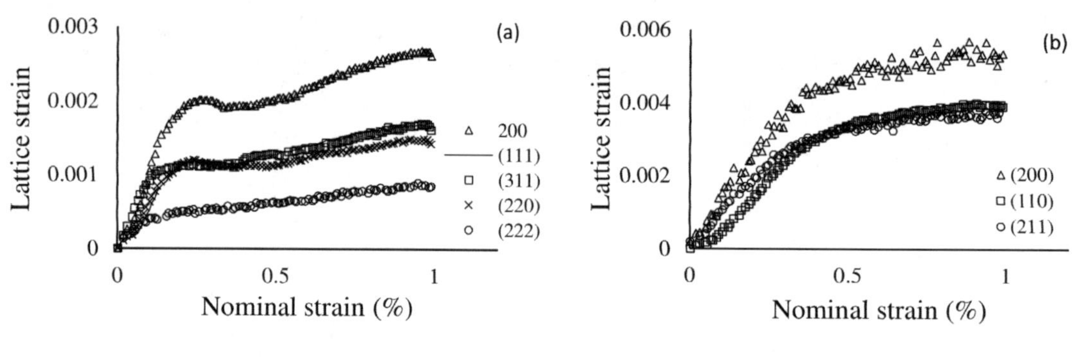

Figure 6. Changes in (a) copper and (b) martensite {hkl} lattice strains as a function of the nominal strain.

Figure 6 shows the lattice strain data plotted as function of applied nominal strain along the loading direction. For the copper phase (figure 6a) it is clearly seen that plastic yielding takes place at around 0.2% strain. In agreement with the data in figure 5, the {222} family shows smaller lattice strains and evidence of plastic yielding at a lower strain. Figure 6b displays the evolution of lattice plane strains of {200}, {110} and {211} in the martensite with nominal strain along the loading direction. The {200} family displays the highest lattice strain while the two other families show a lower lattice strain, again reflecting the elastic anisotropy factor of the cubic crystallographic structure. All three sets of lattice planes show a deviation from linearity (figure 6b) at around a strain of 0.4%, indicating the onset of plastic flow in the martensite phase. The lattice strains measured of the martensite reflections are all larger than those for the copper phase. The data show that for both phases the lattice strains exhibit two different stages of deformation. The first linear region corresponds to the elastic loading in both the copper and martensite. With increasing applied strain, the copper matrix starts to yield, while the martensite continues to undergo elastic deformation. The difference of lattice strain in copper and martensite indicates that load partitioning takes place after yielding of the copper. With further loading the lattice strain of the copper phase shows a gradual increase while the lattice strain in the martensite continues to increase linearly until around 0.4% applied strain, at which point plastic yielding in the martensite takes place. A limited work hardening can be seen in martensite phase, whereas copper phase shows work hardening throughout the entire loading.

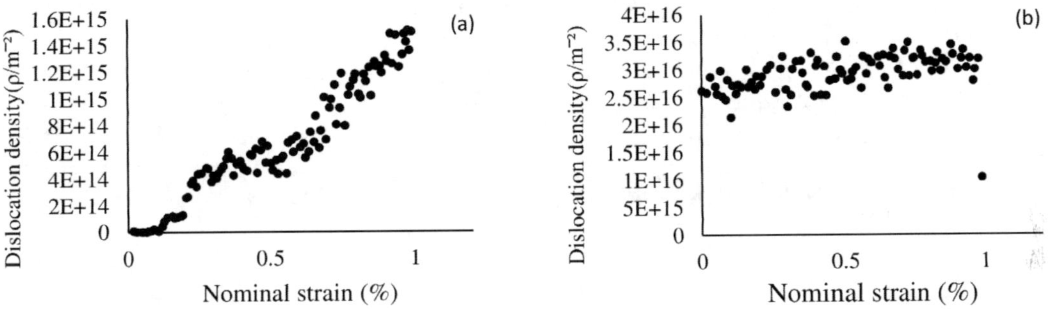

Figure 7. Dislocation density evolution in (a) copper and (b) martensite as a function of the nominal strain.

The pattern of load distribution between the two phases are also seen in the calculated changes in dislocation density with increasing nominal strain during tensile deformation, illustrated in figure 7. Before loading, the martensite phase already contains a moderate dislocation density of about $2.6 \times 10^{16}\,m^{-2}$ compared to a value of $9.5 \times 10^{12}\,m^{-2}$ in the copper phase. As the loading begins, extending from strains of 0% to about 0.11%, which is the region of elastic deformation, the dislocation density shows no change in copper phase. After it has entered the fully plastic flow stage, the dislocation density increases continuously with increasing strain up to $1.5\times10^{15}\,m^{-2}$ before unloading, showing work hardening in the copper phase. In contrast, the dislocation density of the martensite phase shows a wide scatter and increases only slightly throughout the investigated load range, with a maximum dislocation density in the martensite during loading of $3.5 \times 10^{16}\,m^{-2}$.

4. Summary

The micromechanical behavior of a modal dual phase steel and the relationship between microstructure and mechanical properties of its two phases have been investigated by synchrotron X-ray diffraction during tensile deformation at room temperature. The use of a model dual phase system represents a novel approach for fundamental studies of the deformation behavior of dual phase steels as a full separation of the different peaks from the hard martensite and soft copper in X-ray diffraction measurements provides data on the load transfer between the two phases without the need for peak deconvolution methods and related assumptions about the diffraction peak shapes. Moreover, the use of powder mixing and sintering allows independent control of both the size and volume fraction (and potentially the hardness) of each phase, thereby facilitating exploration of optimized microstructural combinations.

At low applied stresses, elastic loading of both phases was found to occur. At higher stresses, macroscopic yielding of the copper matrix took place, the load partitioning among phases was observed and the hard martensite bore significantly larger load than the soft copper matrix. In the copper matrix, the {222} family showed the lowest lattice plane strain and earlier yielding. Both phases had some dislocations before loading, but it is much higher in martensite and increased slightly throughout the loading process. The dislocation density in the copper phase started to rise after entering the plastic stage and increased sharply before unloading showing work hardening in this phase while the martensite phase has just entered the plastic deformation region.

5. References

[1] Mahesh K, Sankaran S and Venugopal P 2012 *Mater. Sci. Technol.* **28:12** 1085-94
[2] Ramazani A, Kazemiabnavi S and Larson R 2016 *Acta Mater.* **116** 231-37
[3] Singh M, Das A, Venu Gopalan T, Mukherjee K, Walunj M, Nanda T and Ravi Kumar B 2018 *Metal. Mater. Trans.* A **49** 463-75
[4] Xu X, Van der Zwaag S, Xu W 2016 *Wear* **348-9** 80-8
[5] Gauss C, Souza Filho I R, Sandim M J R, Suzuki P A, Ramirez A J and Sandim H R Z 2016 *Mater. Sci. Eng.* A **651** 507-16
[6] Jimenez-Melero E, van Dijk N H, Zhao L, Sietsma J, Wright J P and van der Zwaag S 2011 *Mater. Sci. Eng.* A **528** 6407-16
[7] Miao Y, Mo K, Zhou Z, Liu X, Lan K C, Zhang G, Miller M K, Powers K A, Almer J and Stubbins J F 2015 *Mater. Sci. Eng.* A **625** 146-52
[8] Cong Z H, JIA N, Sun X, Ren Y, Almer J and Wang Y D 2009 *Metal. Mater. Trans.* A **40** 1383- 87
[9] Choi K S, Liu W N, Sun X, Khaleel M A, Ren Y and Wang Y D 2008 *Metal. Mater. Trans.* A **39** 3089-96
[10] Wang L, Li M and Almer J 2014 *Acta Mater.* **62** 239-49
[11] Fu B, Yang W Y, Wang Y D, Li L F, Sun Z Q and Ren Y 2014 *Acta Mater.* **76** 342-54
[12] Adachi H, Karamatsu Y, Nakayama S, Miyazawa T, Sato M and Yamasaki T 2016 *Mater. Trans.* **57:9** 1447-53
[13] Shi R, Nie Z, Fan Q, Wang F, Zhou Y and Liu X 2018 *Mater. Sci. Eng.* A **715** 101-07
[14] Song C, Yu H, Lu J, Zhou T and Yang S 2018 *Mater. Sci. Eng.* A **726** 1-9
[15] Zhang M, Qing Tana, Ding J, Chen H, Guo F, Yang Ren and Wang Y D 2018 *Mater. Sci. Eng.* A **729** 444-51
[16] Woo W, Em V T, Kim E Y, Han S H, Han Y S and Choi S H 2012 *Acta Mater.* **60** 6972-81
[17] Yan K, Liss K D, Timokhina I B and Pereloma E V 2016 *Mater. Sci. Eng.* A **662** 185-97
[18] Jia N, Cong Z H, Sun X, Cheng S, Nie Z H, Ren Y, Liaw P K and Wang Y D 2009 *Acta Mater.* **57** 3965-77
[19] Alexander M Korsunsky, Karen E Wells and Philip J Withers *Scr. Mater.* **39:12** 1705-12
[20] Williamson G K and Smallman R E 1956 *Phil. Mag.* **1:1** 34-46
[21] Smallman R E and Westmacott K H 1957 *Phil. Mag.* **2:17** 669-83

Acknowledgements
Financial support from the National Natural Science Foundation of China (project number 5167010359) is gratefully acknowledged.

IOP Conf. Series: Materials Science and Engineering **895** (2020) 012003 doi:10.1088/1757-899X/895/1/012003

Evaluation of Surface Roughness and Cutting Temperature of EN31 Steel with Varying MQSL Parameters

Sai Kiran Chary Nalband[1], Kalyan Pamidimukkala[1], Prathik Akkapally[1], Muralidhar Reddy Challa[2], Yeshwanth Prasanna Kirupakaran[1], Hitesh Dammu[1], Ajay Thati[1], Sachin Jacob Abraham[1], Rakesh Kumar Gunda[1]*

[1]Mahatma Gandhi Institute of Technology, Telangana, India-500075
[2]Indian Institute of Technology Hyderabad, Telangana, India-502285

kumarrakeshgunda@gmail.com

Abstract. In the present-day manufacturing industry, achieving machining accuracy, high production rate, high material removal rate and increasing tool cutting life is becoming more vital and a particularly difficult task. Machining of hard material imposes a lot of challenges, mainly high temperature generation between tool-workpiece interface. Enormous heat generation adversely affects surface roughness, tool life, dimensional quality of the workpiece and deteriorates the tool life significantly. The use of proper lubrication in machining processes is critically important to modify the conditions of contact area by effective control over frictional interaction which should ensure change in tool-chip interface and machining mechanics. In the present work an experimental investigation was carried out on various MQSL parameters on surface roughness and tool temperature in turning EN 31 steel and the results are compared with dry turning. In order to apply lubricants at shearing interface (chip flow at the rake face of the tool) high velocity MQSL set-up is fabricated. The present work has achieved significant results and helped in concluding the influence of lubricant parameters of MQSL in turning process.

1. Introduction

Since the past two decades the practicality of conventional lubrication techniques like flood lubrication are being reconsidered due to the involvement of socio, economic and ecological aspects. There is an emerging need for green and preservative manufacturing. The recycling of the lubricant being difficult and high-priced, [1] concludes that in flood lubrication the total liquid costs (including amount spent on refrigeration) amounts for 17% of the total production cost. The National Institute for Occupational Safety and Health (NIOSH) assesses that about 1.2 million workers are involved in cutting operations and other metalworking operations are exposed to metalworking fluids every year [2]. Exposure of skin to such liquids represent health related concerns, as does the inhalation of the evaporated and other aerosol particles causes pulmonary problems and several types of cancer [3,4]. This makes the use of cutting fluids an issue based on both environmental and health concerns both long-term and short-term consequences.

However, the global demand for productivity and better quality products is constantly increasing by the day. This paired along with the rise of competition in the industrial sector and strong requirement to reduce the cost, the need for high material removal rate along with stability as well as long life of cutting tool is more than ever. In comparison with conventional cutting, machining with high speeds increases the efficiency, the production rate, the accuracy and the overall quality of the final workpiece. These processes generate excessive cutting temperature zone due to the high velocity and feed rate which often affect the surface roughness. The cutting tool will be softened and the surface roughness of the workpiece will be deteriorated [5]. Roughness plays a vital role in forecasting the performance of the machine members as cracks lead to undesirable effects like corrosion and failure. Surface roughness affects the fitting between mating components, ability to withstand corrosion, flow of liquids along pipes, aesthetic appeal of a product. It has become increasingly important to provide appropriate cutting conditions to achieve better surface finish. The need for better quality machining put together with the downsides of using conventional lubrication make alternative lubrication techniques all the more important [6,7].

MQL on the other hand meets all the aforementioned requirements. As the name suggests, the quantity of applied lubricant is reduced impressively to often 50 ml/hour-500 ml/hour, as compared to the sizable thousands of litres of coolant per hour in conventional flood lubrication. This offers other advantages like dry chips as well as workpiece after machining. A lot of research is being done on dry and near- dry machining recently, and minimum quantity lubrication (MQL) machining in particular has been accepted as a successful semi-dry application because of its environmentally friendly characteristics. MQL has been studied in many machining processes, such as drilling [8], milling [9], turning [10], and grinding. These studies showed that with the proper selection of MQL system and cutting parameters, it is possible to obtain performance comparable or better than that with flood lubrication. Machining of 6061 aluminium alloy with MQL, dry and flooded lubricant conditions by P S Sreejith [11] studied the effect of dry machining, minimum quantity of lubricant (MQL), and flooded coolant conditions with respect to the cutting forces, surface roughness of the machined work-piece and tool wear and found that MQL condition will be a very good alternative to flooded coolant/lubricant conditions. Therefore, it appears that if MQL properly employed can replace the flooded coolant/lubricant environment which is presently employed in most of the cutting/machining applications, thereby not only the machining will be environmental friendly but also will improve the machinability characteristics.

Effect on tool wear A Attanasio [12] worked on the results obtained from turning tests and SEM analysis of tools, at two feed rates and two cutting lengths, using MQL on the rake and flank of the tool. The results obtained portrays that when MQL is applied to the tool rake, tool life is generally no different from dry conditions, but MQL applied to the tool flank can increase tool life. Experimental measurement and modelling of surface roughness in turning with CaF2 solid lubricant assisted minimum quantity lubrication by Mayurkumar [13] made correlation between the cutting parameters and measured surface roughness in wet and solid lubricant assisted machining and indicate that surface roughness is affected by feed, depth of cut followed by cutting speed. These outcomes can be used by the machining industry to use the best combination of parameters affecting surface roughness with the MQSL approach. The performance of graphite and boric acid powder blended in SAE 40 oil has been assessed in turning by Krishna and Rao (2008)[14]. The workpiece chose was made of EN8 steel and the tool was cemented carbide. Graphite and boric acid powder in various extents by weight, blended in with SAE 40 oil. The results concluded that solid lubricant assisted machining better than a dry and wet machining environment and 20% of boric acid in SAE 40 oil assisted machining, providing better results. Parametric investigation of turning process on en-31 steel by L B Abhanga [10] deals with study the effect of machining parameters such as cutting speed, feed rate, depth of cut, tool nose radius and lubricant on surface roughness while turning En-31 steel. Concluded that the surface roughness increases with increase in feed rate and depth of cut but surface roughness decreases with increase in cutting speed and tool nose radius respectively.

Performance assessment of MQSL: Minimum quantity solid lubrication during turning of Inconel 718 by A Marques, [15] studied the effect of applying solid lubricant (molybdenum disulphide and graphite) mixed with oil, during turning of Inconel 718 using cemented carbide tools in which the concentration of the solid lubricant in the fluid and the flow rate of the mixture were varied to analyse the main output parameters such as surface roughness, cutting forces and tool life and concluded that this to be a cost-effective and environmental-friendly lubrication technique than flood coolant and

sprayed oil with or without graphite to retard all types of damaging processes and to improve machinability characteristics of Inconel 718.
The above-mentioned studies indicate that the application of solid lubricant–assisted machining is a new technique to control the machining zone temperature without polluting the environment and the great potential of using solid lubricants for low-cost machining under different conditions.
A lot of parametric studies have been done by varying MQSL cutting parameters in turning but there is very little literature available when it comes to MQSL parameters. The main goal of this work is to carry out an experimental investigation on various MQSL parameters (position, distance, feed rate) on surface roughness and tool temperature and compare the results with MQL and dry turning of EN31 steel. The summarized effects and results for various MQL parameters are shown in Table 1.

Table 1. Literature review for MQL parameters.

S. No	Author	Type of lubrication	Parameters	Effect on	Results obtained
1	A.Marques, Suresh Kumar, AR Machado	MQL(Molybdenum disulphide and graphite)	Concentration of solid lubricant and flow rate	Surface roughness, cutting forces and tool life	Cost effective and environment friendly than flood lubrication
2	Mayur Kumar A, Kaushik M Patel	MQL (CaF_2 + SAE 40) Cutting fluid	Feed, depth of cut and cutting speed	Surface roughness	Surface roughness is affected by all 3 parameters
3	Rakesh Kumar Gunda, Narala Suresh Reddy	HP- MQL	High speed machining conditions	Tool wear, surface roughness	HP-MQSL at small and constant flow allows better penetration
4	A Attanasio M.Gelfi	MQL(tool wear effects)	Only 2 feed rates and 2 cutting lengths fixed	MQL on tool rake and tool flank	On tool rake, no difference but on tool flank , life increases
5	Karnik, Paulo Davi M, V.N. Gaitonde	Optimum MQL	Cutting speed, feed rate	Surface roughness and specific cutting force	MQL=200 ml/he, cutting speed 200m/min, Feed rate=0.05min/rev
6	P.S. Sreejith	MQL, dry and flood lubrication	High speed machining conditions	Cutting force and surface roughness	MQL is better when compared to wet and dry

2. Experimental Setup and Procedure

The experimental setup (shown in Figure 1) consists of an air-compressor which is connected to a T-joint to equally divide the air flow into the nozzle and the hydraulic reservoir. The T-joint also houses an air flow sensor which is connected to a node MCU. The instantaneous and average flow rate is displayed on a computer connected to the node MCU. A 1.5l air-tight container made with reinforced plastic is used as the reservoir and it has one inlet port for high pressure air, a pipe which is dipped till the bottom end of the reservoir and an outlet port which is connected to this pipe. When air is pumped into the container, it is pressurised and the lubricant liquid is forced to exit from the outlet port with high pressure. This lubricant outlet is connected to a flow control valve to control the amount of fluid flow per unit time and the other end of the flow control valve is connected to the nozzle. The other end of the T-joint is connected to the 2nd inlet port of the nozzle. The nozzle is fitted on a nozzle stand using a T-clamp which is fixed to the movable carriage so that the nozzle will move with the tool. The nozzle stand is moved to perfectly place the nozzle end near the machining zone perpendicular to the tool. The EN31 steel which is to be machined is placed in the 3-jaw chuck in the head stock while the carbide tip tool is fitted into the tool post. IR temperature sensor is used to measure the instantaneous temperature of the machining zone and the tool. The air compressor is turned on and it maintains the air pressure at 4 bar and the flow control valve is set at 182 ml/hour. The spindle is made to rotate at

598 RPM so that the velocity of cut is maintained at 122 m/min. A depth of cut given is 0.5mm is given. Feed is given manually to the tool and turning operation is performed on the EN31 steel workpiece.

Figure 1. MQSL set-up.

3. Result and Discussions

In EN31 steel turning process employing MoS_2 mixed with SAE-20W40 base lubricating oil as a minimum quantity solid lubricant, is a possible environmentally friendly, alternative for effective control of chip-tool interface temperature. Hence in the present evaluation, varying concentrations of MoS_2 with SAE-20W40 base oil has been used as a minimum quantity solid lubricant to provide the proper lubrication and reduce the friction between chip-tool interface and thereby reducing heat generation between the chip-tool and tool work piece interface.

3.1 Surface Roughness:

Surface roughness (Arithmetic Average: Ra): Good surface finish is usually achieved by finishing methods like surface grinding but sometimes it is left to machining. Even if it is to be finally finished by grinding, machining prior to that needs to be done taking into consideration the surface roughness so as to reduce the costs of grinding operation and reduce initial surface defects as far as possible. Built-up edges, feed marks due to irregular feed, irregular deformation of the tool tip are the key factors to avoid. Surface roughness grows sharply in dry machining due to the high rise in temperature and softening of the tip. The MQSL setup succeeded in reducing the surface roughness. Factors of the MQSL system i.e., air pressure, nozzle distance, flow rate of the lubricant, concentration of MoS_2 in the lubricant base oil were varied and results were observed. Due to increase in the air pressure the chips are removed as soon as the turning operation is done resulting in the better surface finish and it helps in avoiding the formation of built up edge resulting in low surface roughness. As the nozzle distance decreases it was observed that more concentrated application of lubricant and air due to which surface roughness is improved. Flow rate of the lubricant is maintained in optimum conditions ranging from 100-300 ml per hour to obtain lower surface roughness. During machining under such high cutting temperature, the solid lubricant forms a thin film on rake face of the tool. The Ra is calculated as the roughness average of a surface measured microscopic peaks and valleys, it is one of the most effective surface roughness measures commonly adapted in general engineering practice and it gives a good general description of the height variations in the surface. Figure 2 shows the average results of Ra (surface roughness) obtained under different parameters. It has been observed from Figure 2 that the Ra was not affected significantly under the considered parameters after analysing the standard deviation bars. However, based on the average values, the surface finish is better under these conditions. The average values of Ra when using different concentrations of MoS_2 in the base oil were the lowest among the other tested parameters. The addition of MoS_2 with MQL yields lower surface roughness. A broad comparison of the surface roughness values in all the three conditions was also documented.

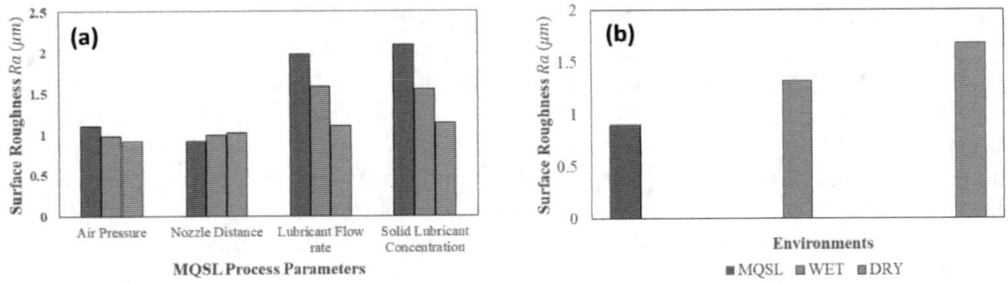

Figure 2. (a) Surface rough measurement during turning under various MQSL process parameters (b) Surface roughness measurement during turning EN31 steel under MQSL, Wet, Dry environment condition.

3.2 Cutting Temperature

The main heat generation zones while machining are, (i) initial zone because of shear and plastic deformation, (ii) chip–tool interface due to deformation and sliding and (iii) work–tool interfaces due to friction. Among those sources substantial effect is done by both the chip-tool and the work-tool interface which mainly influence the manner of chip formation, cutting forces and tool life. In the conducted experiments, efforts were made to decrease the effective cutting temperature. The cutting temperature as shown in Figure 3 taking 3 different values of all input parameters. Conventional flood lubrication fails to provide the required efficiency at both the important interfaces, the chips are still continuous to an extent and these hefty chips make contact with the tool rake surface. On the other hand, it was observed that the MQSL setup aided reduction of the average cutting temperature by about 20–25% depending upon the levels of the process parameters. In the system that was developed the air pressure increases the air and passes through the nozzle and then through the orifice where there is a huge pressure difference because of which the high pressure air condenses due to Joule Thompson effect and the condensed air thus reduces the cutting temperature on a large scale. The factors of the MQSL system like air pressure, nozzle distance, flow rate of the lubricant, concentration of MoS2 in the lubricant base oil were varied. Results helped in concluding that the air pressure has a key role in reducing the cutting temperature. Increase in air pressure also decreases the lubricant particles size and thus resulting in uniform and efficient lubrication. Nozzle distance affects the cutting temperature by varying the area of application, by minimizing the area of application of lubricant, the cooling effect of the lubricant and the condensed air will decrease the heat generated at the cutting interface. Flow rate of the lubricant determines the output amount of lubricant. By lowering the flow rate the temperature also decreased. MoS2 was added in different percentages to the lubricating oil. By varying the concentration of MoS2 in the base oil the thickness of the lubricating film has increased and hence reduce the friction between the work tool interface during turning operation.

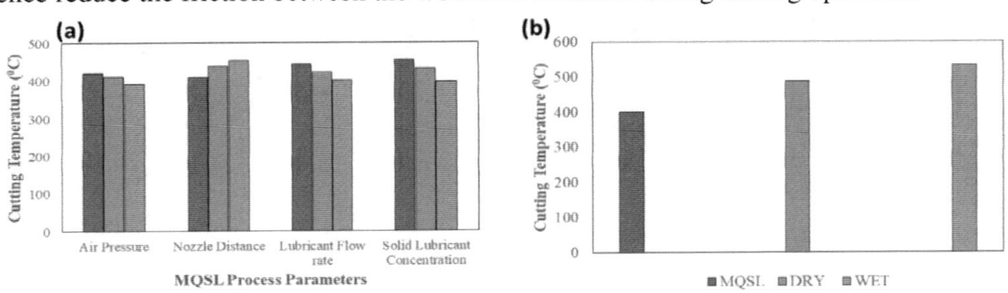

Figure 3. (a) Temperature measurement during turning under various MQSL process parameters (b) Temperature measurement during turning EN31 steel under MQSL, Wet, Dry environment condition.

4. Conclusion

MQSL experimental set-up aiming to improve the overall surface finish; the formation of thin film of lubrication on the rake face of the tool reduces the frictional forces between the tool and chip, which subsequently reduces the temperature developed and prevents the tool wear which further leads to better surface finish. If the fluids are applied precisely to the machining zone, improved results can be expected. Within the selected machining experimental domain, the influence of MQSL in terms of different parameters such as air pressure, nozzle distance, flow rate of the lubricant, concentration of MoS2 in the lubricant base oil, surface roughness of machined surface has been demonstrated during machining of EN31 steel and MQSL. The results demonstrate that the effectiveness of the use of MQSL in reducing friction and heat generation at work–tool interface and contribute to the decrease in the tool wear. Surface roughness and Cutting temperatures of the machined part showed significant variation in different MQSL parameters.

5. References

[1] Kurgin S, Dasch, J, Simon D, Barber G, Zou Q, & Wang X (2009). Environmentally-friendly machining temperature analysis for reaming with minimum quantity lubrication. In Society of Tribologists and Lubrication Engineers Annual Meeting and Exhibition 2009 (pp. 434–435).

[2] National Institute for Occupational Safety and Health (NIOSH). (1998). Criteria for a Recommended Standard–Occupational Exposure to Metalworking Fluids. NIOSH, Cincinnati, OH.

[3] Mackerer C (1989). Health effects of oil mists: a brief review. Toxicology and Industrial Health, 5:429–440.

[4] Thorne P, DeKoster J and Subramanian P (1996). Environmental assessment of aerosols, bioaerosols, and airborne endotoxins in a machining plant. American Industrial Hygiene Association Journal.

[5] Lin W S (2008). Modeling the surface roughness for fine turning of AISI304 stainless steel. In Key Engineering Materials (Vol. 364-366I, pp.644–648).

[6] F. Klocke, G. Eisennbl¨atter, Dry cutting, Ann. CIRP 46 (2) (1997) 519–526.

[7] F. Klocke, G. Eisenblatter, Coated tools for metal cutting-features and applications, Ann. CIRP 48 (2) (1999) 515–525.

[8] Durval U.Bragaa, Anselmo E. Dinizc, Gilberto W.A.Mirandab, Nilvaldo L.Coppinic , Using a Minimum quantity lubricant and a diamond coated tool in the drilling of a aluminum-sillicon alloys.

[9] Alper Uysal, Furkan Demiren, Erhan Altan, Applying MQL method on Milling of Martensitic Stainless Steel by using Nano Mos2 Reinforced Vegetable Cutting Fluid.

[10] L B Abhanga and M Hameedullaha, Parametric investigation of turning process on en-31 steel

[11] P S Sreejith, Machining of 6061 aluminium alloy with MQL, dry and flooded lubricant conditions.

[12] A Attanasio, M Gelfi, Effect on tool wear.

[13] Mayurkumar A Makhesana, Kaushik M Patel, Experimental measurement and modelling of surface roughness in turning with CaF2 solid lubricant assisted minimum quantity lubrication.

[14] Krishna and Rao, The performance of graphite and boric acid powder blended in SAE 40 oil has been assessed in turning (2008).

[15] A Marques, Suresh Kumar Reddy Narala, AR Machado, Rakesh Kumar Gunda, Sravan Kumar Josyula, RB Da Silva, MB Da Silva Performance assessment of MQSL: Minimum quantity solid lubrication during turning of Inconel 718.

Experimental study on characteristics of Ti-6Al-4V after plasma spraying

Dai Chunhui, Lao Xingsheng, Liu yong, Wang Wei

No.19, Yangqiaohu Avenue, Wuhan, Hubei, China

ch.dai@163.com

Abstract. The surface modification of Ti-6Al-4V substrate was carried out by plasma spraying method (WC-17%Co). The surface morphology and composition of the coating were analyzed. Then, under the air environment, the surface of the plasma sprayed WC was carried out. Friction test of three pairs of friction materials filled with 15% glass fiber + 5% graphite, PTFE filled with 25% glass fiber, and PTFE filled with 60% tin bronze. The treated material improves the friction characteristics of the substrate. However, due to the large thickness and hardness of plasma spraying, it is difficult to use sandpaper to reduce the surface roughness, which will affect the dimensional accuracy of the parts.

1. Introduction

Wear is the main reason for shortening the life of mechanical equipment, and the spraying technology can modify its surface[1]. Plasma spraying is a method in which a plasma fox is used as a heat source to heat and accelerate the deposition of powder particles on the surface of a substrate, wherein the particle velocity of the supersonic plasma spraying can reach 600 m/s[2]. Plasma spraying has the advantages of short processing time and high coating quality, and is widely used in various types of equipment[3]to improve the performance of the material[4]. Researchers have carried out more researches on this, such as using TiAl pre-alloy powder plasma spraying to prepare alloy coatings[5,6], using Ti and Al elemental powder plasma spraying to prepare Ti-Al intermetallic compound coatings[7], using Ti and Al ball milling powders to spray coating, and study the reaction and spray products during spraying[8].

The microstructure of the surface modified layer was observed and analyzed by scanning electron microscopy (JEOL JSM-6480). Figure 1 shows the surface topography of the modified sample.

It can be seen from Fig. 1 that the surface modified layer prepared by plasma spraying is relatively rough, and the sprayed particles are agglomerated together in a spherical shape, and there are voids between the particles; this is because the sprayed particles form pores in the process of overlapping each other, and the pores are The subsequently sprayed molten particles are covered and embedded to ensure a high bonding strength between the coatings; in addition, there is a partial melting zone on the surface of the coating; from the energy spectrum, the surface layer is mainly composed of W, C, Co and O elements. Composition, W, C and Co are the main components of the coating, and O element is formed by slight oxidative decarburization of WC particles during coating preparation.

After the sample was cut, the sample was mounted, polished, cleaned, etc., the thickness of the coated layer was measured by a metallographic microscope (OM) to a thickness of 80 μm.

IOP Conf. Series: Materials Science and Engineering **895** (2020) 012004 doi:10.1088/1757-899X/895/1/012004

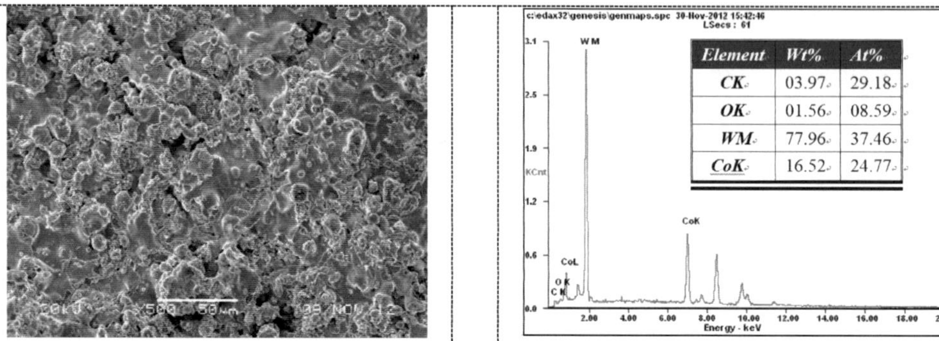

(a) (b)

Figure 1. SEM topography and energy spectrum of coating surface.

2. Friction characteristics experimental scheme

2.1. Friction specimen

The Friction sample parameters are listed in table1, the friction specimen base material is TC4 titanium alloy, and the plasma sprayed tungsten carbide sample can only be ground to 2.0 μm under the existing conditions due to the excessive hardness.

Table 1. Friction sample parameters

material	Sample size	Surface roughness
Substrate	Φ30mm*10mm	≤0.9μm
WC-17%Co	Φ30mm*10mm	2.0μm

2.2. Friction sample

There are three types of friction materials, as shown in Table 2. The end portion having a diameter of 2 mm and a height of 3 mm is a contact end with the disk sample.

Table 2. Friction material

name	material
Material A	Filled with 15% glass fiber and 5% graphite polytetrafluoroethylene
Material B	Filled with 25% glass fiber PTFE
Material C	60% tin bronze PTFE

2.3. Experimental condition

The experiment adopts the pin-disk contact method, the titanium alloy and its surface treatment sample are disc patterns, and the reinforced polytetrafluoroethylene pair is a pin. The wear resistance of the friction pair was evaluated by the HT-1000 friction and wear tester. In the experiment, the pin-disk contact method was adopted, the disk was fixed, and the pin was rotated, as shown in Fig. 2.

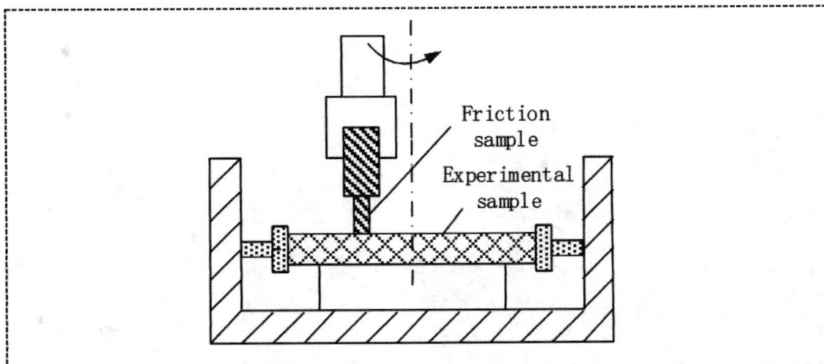

Figure 2. Schematic diagram of friction experiment device.

Wear test trajectory (spin on the disk slide track) Radius: 4mm; disk speed: 336 r / min; contact load: 2000g (20N); wear stroke time: 30min.

3. Results and analysis

3.1. Substrate surface

（1）friction Coefficient

Table 3 shows the friction coefficient of the surface of the substrate and the different reinforcing polytetrafluoroethylene composites. Figure 3 shows the variation of the friction coefficient during the friction process. It can be seen that the friction coefficients of the three reinforced PTFE composites on the surface of the substrate are relatively close, ranging from 0.12 to 0.14.

Table 3. Substrate and different PTFE.

material	Roughness/ m	Change of Sales length/mm	Average friction coefficient
A	0.162	0.997	0.129
B	0.159	1.335	0.140
C	0.158	1.135	0.127

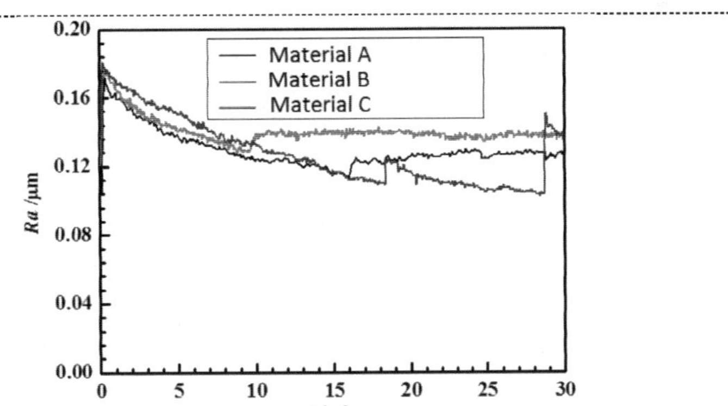

Figure 3. Coefficient of friction between substrate and different materials during rubbing.

（2）Amount of wear

Figure 4 and figure 5 show the wear of the substrate and the different reinforced PTFE materials. It can be seen that the surface of the substrate is worn more seriously and the scratches are more obvious. The damage to the three materials is relatively slight.

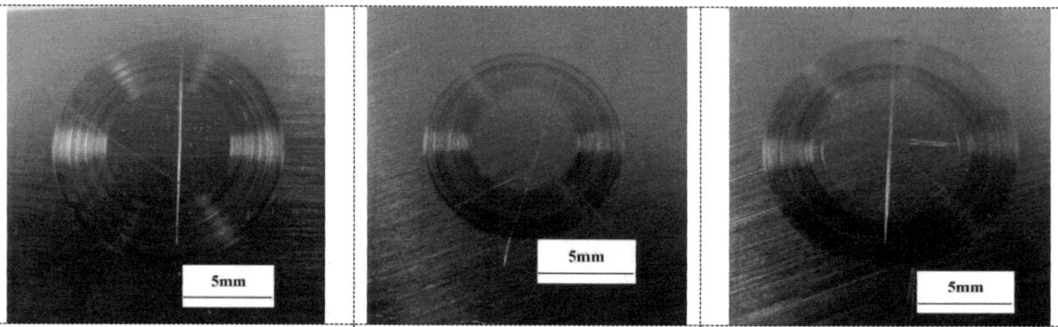

(a) Material A (b) Material B (c) Material C

Figure 4. Substrate wear

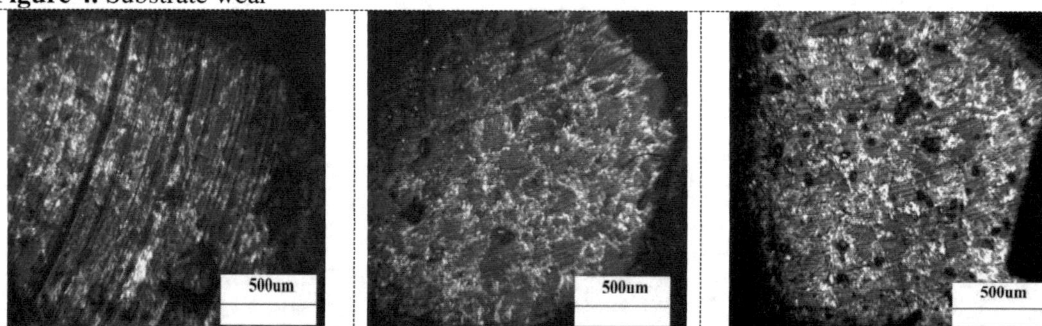

(a) Material A (b) Material B (c) Material C

Figure 5. Enhanced PTFE surface wear.

After testing, the maximum roughness of the surface of the A material is about 1.1 μm, the maximum roughness of the B material is slightly larger, about 2.2 μm, and the maximum roughness of the C material is the largest, reaching 1.6 μm.

3.2. Plasma sprayed WC surface

（1）friction Coefficient

Table 4 shows the friction coefficient of the surface plasma sprayed WC sample and the three friction materials. Figure 6 shows the change of the friction coefficient during the friction process. It can be seen that the friction coefficient of the plasma sprayed WC surface and the three friction materials is between 0.124 and 0.135.

Table 4. Plasma spray WC and different materials.

material	Roughness/ m	Change of Sales length/mm	Average friction coefficient
A	2.663	1.365	0.125
B	2.434	1.591	0.135
C	2.481	1.337	0.124

IOP Conf. Series: Materials Science and Engineering **895** (2020) 012004 doi:10.1088/1757-899X/895/1/012004

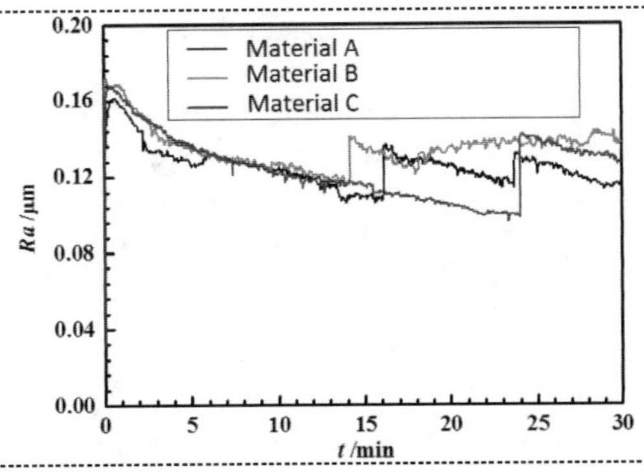

Figure 6. Plasma spraying WC and friction coefficient of different materials.

（2）Amount of wear

Figures 7 and 8 show the wear of the plasma sprayed WC specimen surface and the three mating materials. After testing, the maximum roughness of the A material is about 20.0 μm, the maximum roughness of the B material is relatively large, about 19.5 μm, and the maximum roughness of the C material is about 14.5 μm.

(a) Material A	(b) Material B	(c) Material C

Figure 7. Surface wear of plasma sprayed WC samples.

(a) Material A	(b) Material B	(c) Material C

Figure 8. Enhanced PTFE surface wear.

4. Conclusions

In this paper, the surface modification of Ti-6Al-4V substrate was carried out by plasma spraying method (WC-17%Co). The surface morphology and composition of the coating were analyzed. The thickness of the coating was tested to be 80μm.

IOP Conf. Series: Materials Science and Engineering **895** (2020) 012004 doi:10.1088/1757-899X/895/1/012004

Then, under the air environment, the friction test of the plasma sprayed WC surface and the three pairs of friction materials was carried out. It was found that the friction coefficient was reduced by 0.004, 0.005 and 0.003 compared with the matrix after the surface modification treatment, and The hardness is greatly increased, and the friction characteristics are improved.

On the other hand, plasma spraying also poses certain problems. For example, the thickness of the plasma spray is large, the thickness of the coating affects the dimensional accuracy of the part; after the surface treatment, the hardness increases, resulting in post-processing problems.

5. References

[1] Ferrier W J, Smith K G, Grimm C M. 1999. *Acad Manage J*, vol.42,pp 372-388.

[2] Guessasma S, Montavon G, Coddet C 2005. *Surf Coat Technol*, vol.192, pp70-76.

[3] WU yongzhong, REN Kun,XIE Ou, LU jinbing 2019. *Surf technol*, Vol.48, pp85-90.

[4] Liu S G, Wu J M, Zhang S C, et al. 2007. Wear, Vol.262, pp555-561.

[5] Khor K A, Murakosh Y, Takahashi M, et al.1995. J Mater Process Technol,Vol.48,pp413-419.

[6] Keizoh Honda, Akio Hiroseb, Kojiro F Kobayashibi. 1997. Mater Sci Eng A, Vol.222,pp212-220.

[7] Tsunekawa Y, Gotoh K, Okumiya M, et al.1992. J Therm Spray Technol, Vol.1,pp223-229

[8] Yasuhiro H, Hidekazu M, Kenji M, et al.2002. Mater Sci Eng A, Vol.333:pp92-97.

IOP Conf. Series: Materials Science and Engineering **895** (2020) 012005 doi:10.1088/1757-899X/895/1/012005

Experimental Study on Performance of Anode Humidification System of Large Power Fuel Cell Stack

Lao Xingsheng*, Liu Yong, Dai Chunhui, Wu Jun and Zhao Zhenxing

Science and Technology on Thermal Energy and Power Laboratory, Wuhan 2nd Ship Design and Research Institute, Hubei, 430205, P.R. China

E-mail: 460079620@qq.com

Abstract. The heat and mass transfer characteristics of the hydrogen membrane humidification system of a 70kW atmospheric pressure proton exchange membrane fuel cell stack were experimentally studied. The shell-and-tube membrane humidifier is applied to the high-power fuel cell stack hydrogen humidification system, which has the advantages of fast humidification rate and good wettability of humidified hydrogen. The liquid hydrogen is humidified by liquid water. After the humidification in the experiment, the hydrogen can always reach the supersaturation state. When the hydrogen flow rate is constant, the temperature difference between the membrane humidifier humidification water inlet and outlet decreases with the increase of the water flow rate, and the hydrogen outlet The temperature approaches the humidification water inlet temperature as the liquid water flow rate increases; increasing the humidification water flow rate can reduce the temperature difference before and after the humidification water passes through the membrane humidifier. When the temperature and flow rate of the humidified water are constant, the fuel cell stack load increases, and the relative humidity of the humidified hydrogen outlet does not change significantly.

1. Introduction

Proton exchange membrane fuel cells are characterized by environmental protection, high efficiency, fast startup speed and high power density, and are considered to be the main competitors of future ship power sources. The proton exchange membrane needs to maintain good wettability to ensure the high performance of the fuel cell. The water of the fuel cell is generated at the cathode, and the water of the anode is diffused toward the cathode by electromigration. The hydrogen of the fuel cell needs to be humidified to prevent dehydration of the proton exchange membrane, thereby improving battery performance and working life. Fuel cell stack hydrogen humidification system management is one of the key components of fuel cell system hydrothermal management.

Current fuel cell humidification technologies mainly include self-humidification, internal humidification and external humidification. The external humidification technology is easy to control, easy to install and maintain, and is often used in high power fuel cell stack humidification systems. The commonly used hydrogen humidification system uses a membrane humidification method. The proton exchange membrane fuel cell has a high hydrogen utilization rate, and the water content after the reaction is low. The hydrogen recovered is not enough to humidify the hydrogen inlet, and the liquid water is usually used to humidify the hydrogen. The current research on fuel cell humidification system mainly focuses on the humidity characteristics of fuel cell or low-power fuel cell stack and the matching of humidification technology and stack characteristics [1-5]. The thermodynamic analysis of

IOP Conf. Series: Materials Science and Engineering **895** (2020) 012005 doi:10.1088/1757-899X/895/1/012005

high-power fuel cell stack humidification system is less, directly research on the humidification performance of fuel cell stack humidifiers is still rare.
In this paper, the thermodynamic experimental analysis of the atmospheric pressure fuel cell stack hydrogen humidification system using membrane humidifier is carried out, and the parameter design basis is provided for the hydrogen humidification system management.

2. Membrane humidifier structure and its humidification principle

The membrane humidifier in this paper is a shell-and-tube structure, as shown in figure 1. The tube side passes through the hydrogen, and the shell side passes through the humidifying water. The flow direction of hydrogen and humidifying water is a countercurrent mode.

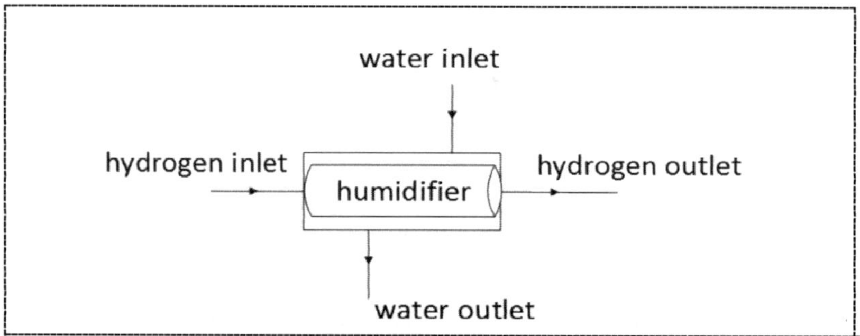

Figure 1. Schematic structure of membrane humidifier.

Membrane humidification is a process of mass transfer and heat transfer, which is a dissolution-diffusion-evaporation model. For the atmospheric system, at the membrane/water interface side, liquid water is dissolved in the surface layer of the membrane, and the water is transferred by the concentration diffusion as the main driving force; at the membrane/gas interface side, the liquid water is evaporated into the gas. Humidification function. The mathematical model can be expressed as follows.
(1) On the membrane/water side, the water concentration on the membrane surface is

$$c_w^{m/w} = \lambda^{m/w} c_s \tag{1}$$

In the formula, the concentration is expressed as mol·cm^{-3}, the subscript w indicates water, and λ the water content of the membrane, h_2o / so_3^{2-} is the number of water molecules per sulfonic acid group in the membrane. The superscript m/w indicates the membrane/water side, and the subscript s indicates sulfonic acid group.
(2) The flux of water in the membrane is

$$N = D_w \frac{dc}{dx} \tag{2}$$

In the formula, N is the water diffusion flux, mol·cm^{-2}·s^{-1}, D is the diffusion coefficient in the film, cm^2·s^{-1}, x indicating the direction along the film thickness, cm, when $\lambda > 4$, there are

$$D_w = 1.25e - 6\exp(2416(\frac{1}{303} - \frac{1}{T})) \tag{3}$$

(3) The heat obtained by humidified hydrogen is

$$\Delta H \ = NAM_{H2} \Delta h \tag{4}$$

In the formula, H is enthalpy, kW, A represents the effective area of the membrane, cm^2, M is the molar mass, kg·mo^{l-1}, and the subscript H_2 indicates hydrogen, h is rate enthalpy, w·kg^{-1}.
(4) The heat released by the humidifying water is

$$Q = c_{p,w} \dot{m}_w \Delta T \tag{5}$$

In the formula, Q means heat, kW, c_p is constant pressure specific heat, kJ·kg-1·K-1, $\dot{m}$ is mass flow rate, kg·s^{-1}, T is temperature, K.

3. Experimental system description

The experimental setup is shown in figure 2. The experimental hydrogen is from a high pressure gas cylinder, and the humidified water is deionized water. After passing through the flow controller, the hydrogen enters the humidifier. The humidified water enters the humidifier through the temperature and flow controller, and the hydrogen and humidified water pressure is controlled by the back pressure regulating valve. There are temperature and pressure sensors on the upstream and downstream of the humidification water side humidifier, and temperature, pressure and humidity sensors on the upstream and downstream of the hydrogen side humidifier. The interval between each two data points in the experiment was 5 minutes.

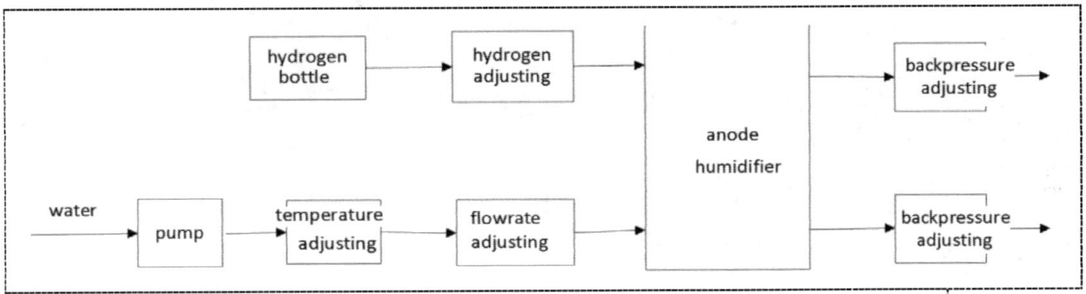

Figure 2. Structure of hydrogen humidifier testing system.

The membrane humidifier parameters are shown in table 1, ρ for density, δ for thickness, subscript m for film, and dry for dry state.

Table 1. parameter values used as the base case.

Parameter	$\rho_{m,dry}$	$M_{m,dry}$	A	δ_m
Value	1.98 kg/cm^2	1100 g/mol	13945 cm^2	0.01524 cm

4. Results and discussion

The membrane humidifier has excellent water permeability. It can obtain good humidification effect by humidifying hydrogen with liquid water. When the humidification water flow is not large, the relative humidity of the humidified hydrogen outlet can reach 100%. The temperature difference between the humidified hydrogen outlet temperature and the humidified water inlet and outlet is the main factor affecting the overall efficiency of the humidification system. When the humidified hydrogen outlet is supersaturated wet hydrogen, the higher the temperature, the less easily the water vapor condenses. When the flow rate of hydrogen is constant, the larger the flow rate of humidified water, the smaller the temperature difference between the inlet and outlet, and the higher the outlet temperature of the humidified hydrogen gas, the greater the consumption of the humidification pump.
During the experiment, the hydrogen inlet temperature was 20^{0}C, the relative humidity was less than 30%, and the humidification water inlet temperature was set to 60^{0}C.
Figure 3 and figure 4 respectively show the relationship between the humidification water flow rate and the temperature and humidity of the humidified hydrogen outlet when the hydrogen flow rate is constant, and the relationship between the humidification water flow rate and the temperature difference between the inlet and outlet. The flow unit is slpm (in the standard state, l/min) and the temperature unit is K (Kelvin). The volume flow rate of hydrogen is 750 slpm, the flow rate of humidification water increases from 10.45 slpm to 18.68 slpm, and the difference between hydrogen

outlet temperature and humidification water inlet temperature is reduced from 2.5K to 0K; the relative humidity of humidified hydrogen outlet does not change much. , always between 120%~130%, because the membrane water activity is much larger than 1 on the membrane/humidified water surface, the water content is much greater than 14, in the atmospheric system, the hydrogen working pressure is low. The evaporation of liquid water on the surface of hydrogen gas has a small evaporation resistance, so the rate of liquid water passing through the membrane to hydrogen is very fast. After passing through the membrane, some liquid water will not desorb from the membrane surface. In figure 4, the humidification water flow rate increased from 10.45 slpm to 18.68 slpm, and the humidification water inlet and outlet temperature difference decreased from 10K to 6.3K. The humidification water flow increased from 10.45 slpm to 18.68 slpm, the humidification pump power will increase significantly, and the humidification hydrogen outlet temperature and humidity conditions will not change much. After entering the fuel cell stack, there will be no significant impact on the performance of the reactor.

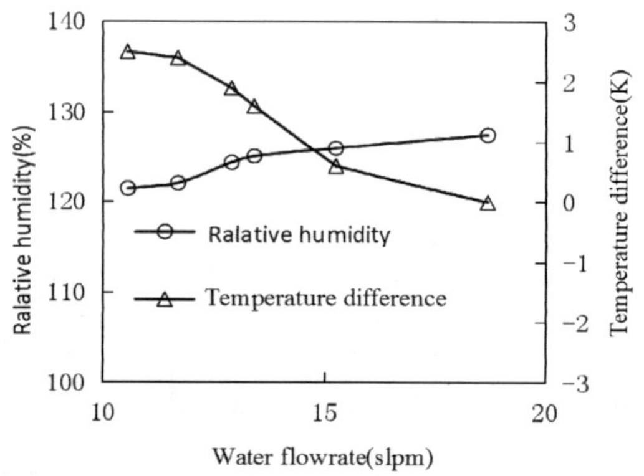

Figure 3. Effect of humidified water flow on hydrogen humidification performance.

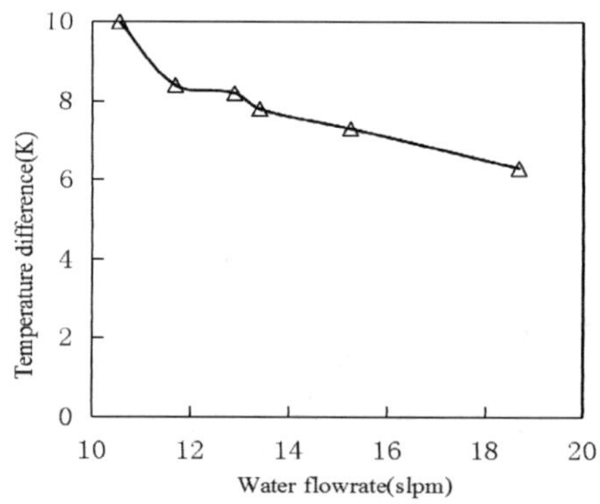

Figure 4. Effect of humidified water flow on temperature difference between humidified water inlet and outlet

Figure 5 and figure 6 respectively show the relationship between the hydrogen flow rate and the temperature and humidity of the outlet when the humidification water flow rate is 12 slpm, and the relationship between the hydrogen flow rate and the temperature difference between the inlet and outlet of the humidified water. The hydrogen flow rate is 300, 500, 750 slpm, respectively. When the hydrogen flow rate increases, the relative humidity of the hydrogen outlet does not change much, between 120% and 130%; the difference between the humidification water inlet temperature and the hydrogen outlet temperature is 0.7K, 1.2K, 2.3K respectively. In Figure 6, the hydrogen flow rate increases from 300 slpm to 750 slpm, and the temperature difference between the humidified water inlet and outlet increases from 4K to 8.4K. When the fuel cell stack load increases, the hydrogen flow rate increases, and the wet hydrogen dew point temperature entering the fuel cell stack does not change much.

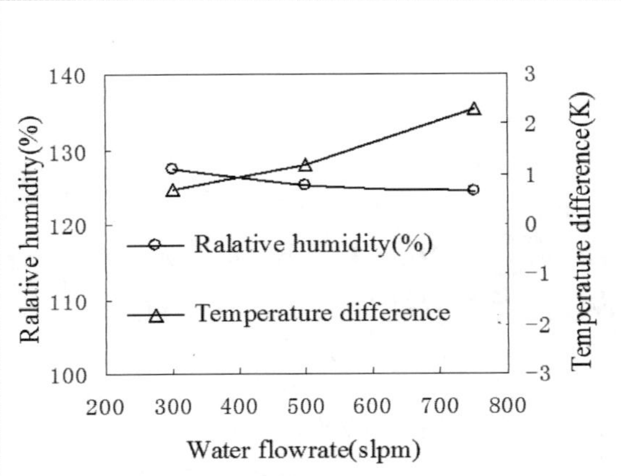

Figure 5. Effect of humidified water flow on anode humidification performance.

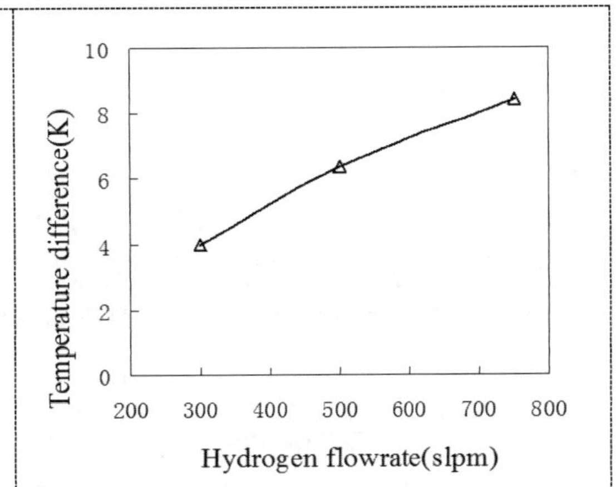

Figure 6. Effect of humidified water flow on temperature difference between humidified water inlet and outlet

According to the equation, the amount of helium gas increase and the amount of humidification water heat energy reduction are calculated separately, and the measurement error is used.

$$\delta_Q = (\Delta H - Q)/[(\Delta H + Q)/2] \quad (6)$$

In the experiment, no more than 10%, the heat balance in the humidification process is basically reflected in the test data, and the experimental data is credible.

5. Conclusions

The experimental study on the heat and mass transfer performance of the hydrogen membrane humidification system of high-power fuel cell stack shows that the membrane humidification system which humidifies hydrogen with liquid water can obtain good hydrogen humidification effect, and the humidified hydrogen outlet is over. Saturated state; hydrogen volume flow rate is 750 slpm, when the ratio of humidified water to hydrogen mass flow is greater than 180, the temperature difference between hydrogen outlet temperature and humidification water inlet temperature is less than 2K when the temperature and flow rate of the humidified water are constant. The relative humidity of the humidified hydrogen outlet does not change significantly as the fuel cell stack load increases.

6. References

[1] Dowon Cha, Wonseok Yang, Yongchan Kim.Performance improvement of self-humidifying PEM fuel cells using water injection at various start-up conditions. 2019 Energy, 183:514-524

[2] Desmond Adair, Martin Jaeger. Quasistatic modelling of PEM fuel cell humidification system. 2018 Materials Today: Proceedings, 5(11) Part 1: 22776-22784

[3] Linhao Fan, Guobin Zhang, Kui Jiao. Characteristics of PEMFC operating at high current density with low external humidification. 2017 Energy Conversion and Managemen, 150:763-774

[4] Tianyao Pan, Jiong Shen, Li Sun, and Kwang Y. Lee. Thermodynamic modelling and intelligent control of fuel cell anode purge. 2019 Applied Thermal Engineering, 154:196-207

[5] Wei-Mon Yan, Chen-Yu Chen, You-kai Jhang, Yu-Hsuan Chang, Mohammad Amani. Performance evaluation of a multi-stage plate-type membrane humidifier for proton exchange membrane fuel cell. 2018, Energy Conversion and Management, 176:23-130.

IOP Conf. Series: Materials Science and Engineering **895** (2020) 012006 doi:10.1088/1757-899X/895/1/012006

Remaining Useful Life Prediction of Lithium-ion Battery based on Attention Mechanism with Positional Encoding

Beitong Zhou[1], Cheng Cheng[1], Guijun Ma[2] and Yong Zhang[3]

[1] School of Artificial Intelligence and Automation, Huazhong University of Science and Technology, Wuhan, 430074, China
[2] School of Mechanical Science and Engineering, Huazhong University of Science and Technology, Wuhan, 430074, China
[3] School of Information Science and Engineering, Wuhan University of Science and Technology, Wuhan, 430081, China

E-mail: zhoubt@hust.edu.cn

Abstract. The rising demands of more reliable and stable electrical systems attach importance to accurate Remaining Useful Life (RUL) prediction of the lithium-ion batteries. As artificial intelligence and machine learning techniques advance, data-driven methods especially deep learning algorithms have become the rising star in RUL prediction. Recurrent Neural Networks (RNNs) and their variants such as Long Short Term Memory have proven effectiveness in various sequential tasks. However, due to its iterative nature along the time axis, RNNs take much time for information to flow through the network for prediction. Inspired by recent advance brought by Transformer in sequence transduction tasks, we proposed the attention mechanism based Convolutional Neural Network (CNN) with positional encoding to tackle this problem. The attention mechanism enables the network to focus on specific parts of sequences and positional encoding injects position information while utilizing the parallelization merits of CNN on GPUs. Empirical experiments show that the proposed approach is both time effective and accurate in battery RUL prediction.

1. Introduction

Due to the high energy density, no memory effect, and low self-discharge traits of lithium-ion batteries, they are becoming widely used in a large amount of energy systems including consumer electronics, electrical vehicles, and aerospace systems [1] [2]. However, with many merits, the batteries can harm the public safety because of a flammable electrolyte and can lead to explosions and unexpected damages. To help fully utilize the advantages of the lithium-ion batteries and prevent them from causing catastrophic damage to human safety, it's necessary to accurately monitor the battery's states. There are several important indicators such as State of Charge (SOC), State of Health (SOH), and Remaining Useful Life (RUL). The lithium-ion battery degrades through repeated charge cycles. The battery capacity, which is defined as the amount of electrical charge a battery can holds in its full state, decreases along with the degradation process. Thus, RUL of a battery indicates the number of remaining charge-discharge cycles before the states degrade to a given failure threshold. For practical use, the risk of causing damage increases when the capacity is reduced to below 70%-80% of its initial value and the battery fails to provide further use [3] [4] [5]. The accurately predicted RUL is of great importance for maintenance in advance and alarming the potential risk before the battery reaches the failure threshold.

Comprehensive research has been conducted on the battery modeling and RUL prediction. Those algorithms can be roughly divided into model-based methods and data-driven methods. The model based methods include equivalent circuit models, electrochemical models, and empirical models. The equivalent circuit model [6] [7] is formed by an open-circuit voltage source connected with electric elements, such as resistors and capacitors. Because the equivalent circuit model is empirical analysis, it depends on impedance data and cannot provide highly accurate prediction results. On the other hand, the electrochemical models [8] [9] reveals the internal physics property of the battery through partial differential equations (PDEs) but it is hard to solve PDEs in constant time. For empirical models, the capacity-based exponential models combined with filter approaches such as Kalman Filter (KF), Particle Filter (PF) and their variants are common solutions. Saha *et al.* [10] used the PF algorithm to predict the battery RUL and the uncertainty representation with the empirical degradation model as the state transition equation. Chang *et al.* [11] proposed a hybrid approach which combines the Unscented Kalman Filter (UKF) method with a double exponential model for long-term capacity prediction. Miao *et al.* [12] obtained the prediction results of the lithium-ion battery RUL by using UKF algorithm also based on the degradation model. Although empirical models are preferred from practical perspective, there are still demands of a more precise RUL prediction.

Data-driven methods, such as Support Vector Regression (SVR), Relevance Vector Machine (RVM), and Box-Cox Transformation (BCT), make predictions in another different way which learn latent representations directly from history data. Dong *et al.* [13] used the Support Vector Regression - Particle Filter (SVR-PF) algorithm to provide RUL prediction as well as SOH monitoring. Zhang *et al.* [14] developed an RUL prediction method based on the BCT and monte carlo simulation.

Recently, as artificial intelligence and deep learning methods prevail in various tasks such as computer vision [15], image caption [16] and neural machine translation [17], Recurrent Neural Network (RNN) and its variants have proven its effectiveness in processing with sequential tasks including RUL predictions. Liu *et al.* [18] proposed an Adaptive Recurrent Neural Network (ARNN) for system dynamic state forecasting and its effectiveness was demonstrated in RUL prediction. Zhang *et al.* [19] employed Long Short-Term Memory (LSTM) to learn the long-term dependencies among the degraded capacities of lithium-ion batteries. Convolutional Neural Network, another popular deep learning network in image classification [20], has already been employed in mechanical systems such as fault diagnosis [21] [22]. Cheng *et al.* [23] proposed an online data-driven framework to exploit the adoption of deep CNN in predicting the RUL of bearings. Ma *et al.* [24] used a hybrid neural network that combines the advantages of CNN and LSTM to make RUL predictions.

Deep learning methods, which extract features from unstructured data automatically and have abilities of fitting complex datasets, have already outperformed model-based methods and traditional machine learning methods in RUL prediction. However, due to the iterative nature of LSTM, it can take a lot of time for information to flow through the network sequentially with long sequences, making it theoretically impossible to exploit the merits of concurrent hardwares. CNN, on the other hand, would not be affected by the length of input sequences and is prone to parallelization as information of each kernel within the same layer is inferred in isolation. A network, which can be implemented in parallel and able to keep position information within the input data, is of great importance to practical use. Being released in late 2017, the Transformer architecture proposed in [17] has refreshed multiple SOTA records in neural machine translation tasks. The so-called Transformer network leverages fully connected networks with a cleverly-designed attention mechanism, which indicates the directions of possible solutions to our above problems.

Inspired by Transformer, this paper proposes to use CNN based on attention mechanism with positional encoding to predict the RUL of lithium-ion batteries. The implementation contains following stages. Firstly, the sliding window approach with a given sequence length will process the raw data and the CNN layer acts as a low dimensional embedding to extract features from inputs. Secondly, position information will be injected into embeddings through positional encoding. Meanwhile, attention weights are computed to help the network focus on specific parts of intermediate information. Thirdly, a fully connected layer is used to achieve the final RUL prediction, where the RUL can be calculated once the predicted capacities reach the failure threshold.

This work makes the following contributions: 1) High accuracy. Empirical experiments are conducted to prove its effectiveness of making precise RUL predictions compared with other methods. 2) Attention mechanism and positional encoding are adopted. Positional encoding injects position information which can make up for the weakness CNN has compared with RNN. Attention mechanism help focus on specific parts and improve the robustness and performance of the network. 3) Easy to implement in parallel. The proposed algorithm overcomes the weakness of RNN when the sequence becomes long and can fully exploit the parallelization ability of GPUs.
The paper's outline is as follow. Section 2 introduces the proposed algorithm framework. In section 3, experiments are conducted and detail analysis is included. Section 4 summarizes the whole work and indicates future works.

2. Algorithms

To solve the above-stated problem, the schematic of proposed framework is shown in Figure 1 which involves three main parts. Related algorithms will be discussed in detail.

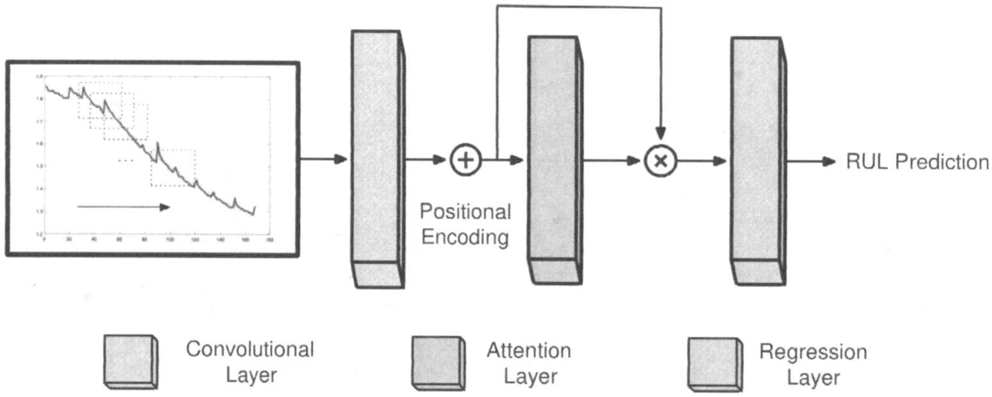

Figure 1. RUL prediction framework, which is mainly comprised of three parts: (1) Convolutional layer; (2) Positional encoding and attention mechanism; (3) Regression layer.

2.1. Convolutional neural network

CNN proposed by Lecun *et al.* in works [25] and [26] becomes the most well-known deep learning architecture since its superior performance in computer vision tasks such as image classification [20] and object detection [27]. A typical CNN architecture mainly contains local connections, shared weights, and local pooling [28].
In this work, only a 1-D CNN layer is used for simplicity, since the input data are time sequences. Convolutional layer usually contains multiple kernels and every kernel has a filter w ∈ Rf and a bias b ∈ R. The output feature v_i is obtained through a convolution operation, which is between an input kernel of size f and the filter w, and a non-linear activation function ReLU to avoid potential vanishing gradient problem. The whole transformation for a single kernel is described as:

$$v_i = ReLU(w * u_j + b) \tag{1}$$

where $w_j \in \mathbb{R}^{1xf}\}$ represents the j-th sub-vector of the input data, $v_i \in \mathbb{R}$ is the i-th value in the convolution outputs and '$*$' denotes the convolution operation. The outputs of the convolution operation $\mathbf{v} = [v_1, v_2, \dots, v_o]$ is also called as the feature map, where $o = (l - f)/s + 1$ is the number of output features, l is the length of input sequence length and s is the stride for convolution. Usually for a typical convolutional layer with k kernels, we would have k independent feature maps stacked together.

2.2. Positional Encoding and Attention Mechanism

Positional encoding, which was first introduced in [17], injects positional information about the relative or absolute position of the sequence. In this paper, the implementation is the same as the original work except that we chose 1000 not 10000 due to lower dimensions and shorter sequence length:

$$PE_{(pos,2i)} = sin\left(pos/1000^{2i/d}\right)$$
$$PE_{(pos,2i+1)} = cos\left(pos/1000^{2i/d}\right) \quad (2)$$

where the pos is the position, i is the index of input dimensions and PE is the abbreviation of positional encoding. We visualized the positional encoding in heatmap in Figure 2. In the figure, each row which is sequentially ordered corresponds to the positional encoding of a vector. Adding this kind of position-dependent information may help the model incorporate the order of each positions.

Attention mechanism was first proposed in the work [29] which integrates attention mechanism with an encoder-decoder architecture on neural machine translation tasks. Many deep learning models in time series tasks are still shallow due to the lack of interpretability and large-scale data while attention mechanism provides a compelling solution. Lai *et al.* [30] proposed LSTNet and was one of the first to combine LSTM with attention to forecast multivariate time series. Shih *et al.* [31] proposed a novel attention mechanism for selection and use frequency domain information for time series forecasting.

The output of positional encoding is $k \in \mathbb{R}^{k\times l}$ and position weights $s \in \mathbb{R}^{k\times l}$ are computed through attention layer. A fully connected layer with weights $w \in \mathbb{R}^{l\times l}$ and a bias $b \in \mathbb{R}^{l}$ is used to implement the modified attention mechanism. The outputs are normalized using the softmax function:

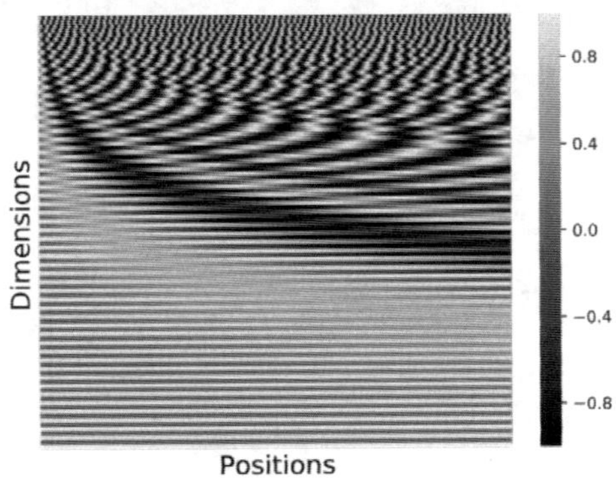

Figure 2. The heatmap visualization of positional encoding. There are 100 rows and each row is a vector of length 500.

$$s_{ij} = \frac{exp(e_{ij})}{\sum_{m=1}^{l} exp(e_{im})} \quad (3)$$

where $e_i = wh_i + b$. The final outputs of the attention layer is the element-wise production between the intermediate representations h and the position weights s.

2.3. Regression Layer

Finally, the outputs of attention layer are fed into the regression layer which is simply a fully connected linear layer without any activation function. For a regression problem, the commonly used mean squared error (MSE) function is chosen as the error metric to evaluate the model performance:

$$e = \frac{1}{N}\sum_{m=1}^{N}(y_m - \widehat{y_m})^2 \quad (4)$$

where y_m is the ground label and $\widehat{y_n}$ is the predicted value. The goal of optimizing the whole network is to minize the mean squared error by optimizers such as Adagrad [32] and Adam [33].

The overall framework of the proposed algorithm can be summarized in Algorithm 1.

Algorithm 1 Proposed algorithm.

Require: The train set D_{train}
Initialize: The weight parameters of convolutional layer, attention layer and the final regression layer
1: **while** not reach the maximum iterations **do**
2: Get degradation data from raw inputs using sliding window approach.
3: Feed degradation data into the convolutional layer and get the low dimensional embeddings v.
4: Add position information using positional approach.
5: Feed intermediate representations h into the attention layer to compute position weights and outputs are the element-wise production of representations h and the position weights s.
6: Use the final regression layer to make RUL predictions and calculate the MSE loss
7: Compute the gradients with Adam optimizer and update the parameters.
8: **end while**

Table 1. Evaluation Metrics.

Metrics	Equations
Error Metric	$EM = \lvert R - \hat{R} \rvert$
Accuracy Metric	$AM = \left(1 - \frac{\lvert R - \hat{R} \rvert}{R}\right) \times 100\%$
Steady Metric	$SM = \sqrt{\frac{1}{n-k} \sum_{i=k+1}^{n} (x_i - \hat{x_i})^2}$

2.4. Evaluation Metrics

We chose three evaluation metrics listed in Table 1 to estimate the model performance. The Error Metric (EM), Accuracy Metric (AM) and Steady Metric (SM) are frequently used metrics for comparisons.

In the table, R and $\hat{R}$ represents the actual RUL which is the number of charge-discharge cycle when SOH reaches the failure threshold and the estimated RUL. For SM, x_i and $\hat{x_i}$ are the actual capacity and the predicted capacity in test set D_{test}. k is the number of cycles in train set D_{train} and the n is the total number of cycles for the dataset. A model which can achieve accurate predictions is supposed to have small enough EM and SM while AM should be large.

3. Experiments

3.1. SOH and RUL

State of Health (SOH) is a commonly used indicator of the battery's operating states in a repeated charge-discharge cycle. SOH can be calculated using capacity, internal resistance, cycle number, etc. For simplicity, the capacity ratio is used to describe SOH. The SOH SOH_k in cycle k is described as:

$$SOH_k = \frac{x_k}{x_0} \times 100\% \quad (5)$$

where x_0 is the initial capacity and x_k is the battery capacity in the number k charge-discharge cycle. The battery RUL can be defined as the number of remaining char-discharge cycles in a certain time based on historical degradation curve.

3.2. Dataset Description

The dataset used in following experiments has been collected from a battery prognostics test bed at the NASA Ames Prognostics Center of Excellence (PCoE). The commercially available Li-ion 18650 sized rechargeable batteries with a Graphite anode and Lithium Nickel Cobalt Manganese oxide (LiNiCoMnO2) cathode are used. The test bed contains a programmable 4-channel Direct Current (DC) electronic load, programmable 4-channel DC power supply, voltmeter, ammeter, etc.. The lithium-ion batteries were run through three different operational profiles (charge, discharge and Electrochemical Impedance Spectroscopy) at different temperatures. The battery charge was carried out in a Constant Current (CC) mode at 1.5 A till the battery voltage reached 4.2 V, the continued in a Constant Voltage (CV) mode until the charge current dropped to 20 mA. Discharges were carried out at different current load levels until the battery voltage fell to preset voltage thresholds. Repeated charge and discharge cycles resulted in accelerated aging of batteries.

3.3. Results

For each lithium-ion battery, we keep using the same hyper-parameters such as kernel size, stride size, and learning rate to guarantee the stability and robustness. The model architecture is shown in Table 2.

Table 2. Model Architecture.

	Input Size	Output Size	Layer Setting
Input Layer	1x1x20	...	...
Convolutional Layer	1x1x20	1x8x16	k=8, f=5, s=1
Positional Encoding	1x8x16	1x8x16	...
Attention Layer	1x8x16	1x8x16	16 neurons
Flatten	1x8x16	1x128	...
Regression Layer	1x128	1x1	1 neuron

Table 3. Comparisons of predicted results between the proposed network and other existing deep learning methods.

	M_{th}	R	CNN				LSTM				CNN + Attention			
			R_b	EM	AM	SM	R_b	EM	AM	SM	R_b	EM	AM	SM
Battery-1	88	73	45	28	61.64%	0.0796	46	27	63.01%	0.0579	72	**1**	**98.6%**	**0.0108**
Battery-2	60	46	61	15	67.39%	0.0369	26	29	56.52%	0.0615	36	**10**	**78.26%**	**0.0217**
Battery-3	119	46	31	15	67.395	0.0304	58	12	73.91%	0.0116	51	**5**	**89.13%**	**0.0057**

For input data, the sliding window size is chosen as 20, corresponds to the input length 1 of convolutional layer. The whole network is optimized using Adam optimizer. The learning rate and weight decay are set as 0.001 and 0.005.

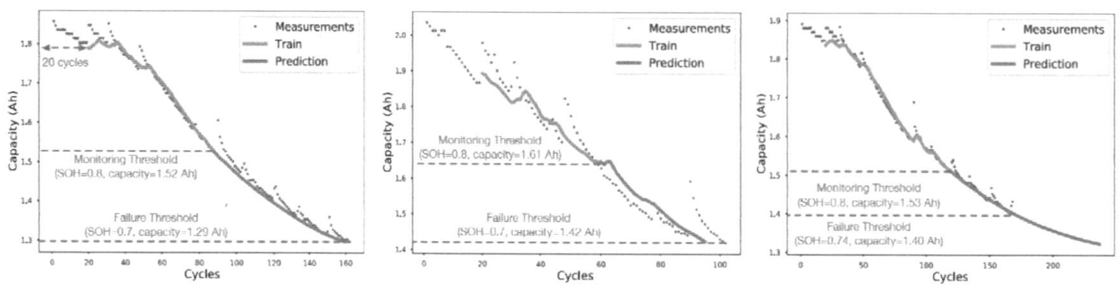

(a) Battery-1 (b) Battery-2 (c) Battery-3

Figure 3. Prediction results of 3 different lithium-ion batteries in the NASA dataset. The sliding window size here is set to 20 so that no predictions are made in first 20 cycles. The monitoring thresholds, failure thresholds and their corresponding capacities are annotated in the figure. Battery-3 didn't reach the failure threshold so that the lowest capacity was chosen as the failure threshold and the cycle of it was the actual RUL.

To conduct experiments to evaluate model performance, a monitoring threshold of 0.8 SOH was chosen to split train set D_{train} and test set D_{test}. The actual RUL was defined as the charge-discharge cycles needed to reach the failure threshold of 0.7 SOH. There are 3 lithium-ion batteries in the NASA dataset and the prediction results are shown in Figure 3.
Then network fits the train set well and predicts the RUL accurately as we can see in the Figure 3. To understand what the attention mechanism is actually doing, we randomly choose a input sequence from the Battery-3 and visualize the position weights in heatmaps. The visualization is shown in Figure 4.

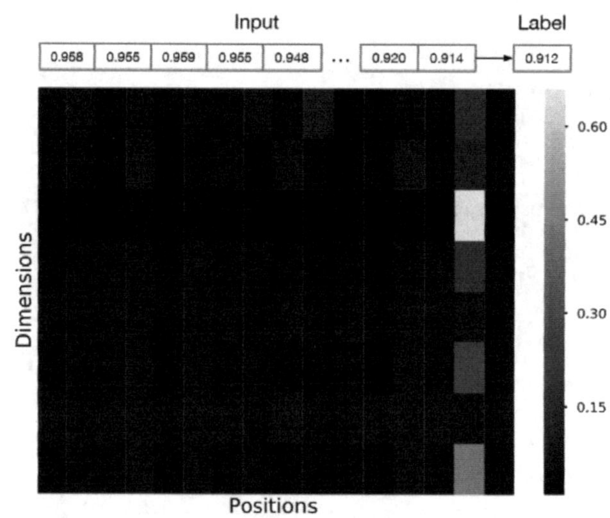

Figure 4. Heatmap visualization for position weights in the attention layer given a random input. The convolutional layer reduces the sequence length from 20 to 16 and the dimensions increase to 8. The lightness in a given position indicates the weights in the attention mechanism.

In the above heatmap, we can notice that the layer pays more attention as the sequence reaches the end and for different dimensions, usually different weights are given. It shows that the attention mechanism can actually capture the position order and focus more on areas which are closer to the prediction label.

3.4. Comparisons
We also compare the proposed algorithm with other CNN and LSTM. The experiment including evaluation metrics results for two methods are listed in the work [24]. The final results are listed in Table 3. As we can see, our method achieves the lowest EM, SM and the highest AM. It shows that compared with other methods, the proposed algorithm predicts the RUL more accurately and steadily.
Another comparison experiment is conducted to verify the time effectiveness of CNN. A stacked LSTM with 2 layers and 8 hidden units is chosen as the comparing object. Randomized inputs with different input lengths 10,50,100,200,500 are fed into two networks and running for 100 times. The average inference time are recorded and the experiments are conducted using Intel(R) Core(TM) i74770HQ CPU in macOS Catalina. The results are shown in Table 4. The increased input length only adds marginal time cost to CNN inference time but for LSTM, the cost can be hard to ignore.

Table 4. Average inference time for two networks.

Model	Input Length				
	10	50	100	200	500
CNN + Attention	0.485ms	2.189ms	2.512ms	2.584ms	4.024ms
Stacked LSTM	2.007ms	45.15ms	85.93ms	175.3ms	409.7ms

4. Conclusion
In this work, we propose a convolutional neural network based on attention mechanism with positional encoding to predict the RUL of lithium-ion batteries. The convolutional layer is used to extract features and translate the inputs into a low dimensional space. The positional encoding injects position information using a encoding function which is relevant to the dimension and sequence position. The attention mechanism helps focus on specific parts of intermediate representations. Finally, a regression layer is used to make RUL predictions. Results on the NASA dataset shows that the proposed network can indeed capture hidden trending and make accurate RUL predictions. We also compare the averaged inference time with LSTM to verify the superior advantage of CNN for practical use.
Future work includes: 1) find out a more reliable way to cross validate the model performance and 2) improve the algorithm to reduce the instability brought by random optimization.

5. References
[1] Liu X, Wu J, Zhang C and Chen Z 2014 *Journal of Power Sources* **270** 151–157
[2] Tarascon J M and Armand M 2011 Issues and challenges facing rechargeable lithium batteries *Materials for Sustainable Energy: A Collection of Peer-Reviewed Research and Review Articles from Nature Publishing Group* (World Scientific) pp 171–179
[3] Liu D, Wang H, Peng Y, Xie W and Liao H 2013 *Energies* **6** 3654–3668
[4] Tang S, Yu C, Wang X, Guo X and Si X 2014 *Energies* **7** 520–547
[5] Dubarry M and Liaw B Y 2009 *Journal of Power Sources* **194** 541–549
[6] Hu X, Li S and Peng H 2012 *Journal of Power Sources* **198** 359–367
[7] Remmlinger J, Buchholz M, Meiler M, Bernreuter P and Dietmayer K 2011 *Journal of Power Sources* **196** 5357–5363
[8] Zhang W J 2011 *Journal of Power Sources* **196** 13–24
[9] Kemper P, Li S E and Kum D 2015 *Journal of Power Sources* **286** 510–525
[10] Saha B, Goebel K and Christophersen J 2009 *Transactions of the Institute of Measurement and Control* **31** 293–308
[11] Chang Y, Fang H and Zhang Y 2017 *Applied energy* **206** 1564–1578
[12] Miao Q, Xie L, Cui H, Liang W and Pecht M 2013 *Microelectronics Reliability* **53** 805–810
[13] Dong H, Jin X, Lou Y and Wang C 2014 *Journal of power sources* **271** 114–123

[14] Zhang Y, Xiong R, He H and Pecht M G 2018 *IEEE Transactions on Industrial Electronics* **66** 1585–1597
[15] He K, Zhang X, Ren S and Sun J 2016 Deep residual learning for image recognition *Proceedings of the IEEE conference on computer vision and pattern recognition* pp 770–778
[16] Jia X, Gavves E, Fernando B and Tuytelaars T 2015 Guiding the long-short term memory model for image caption generation *Proceedings of the IEEE international conference on computer vision* pp 2407–2415
[17] Vaswani A, Shazeer N, Parmar N, Uszkoreit J, Jones L, Gomez A N, Kaiser Ł and Polosukhin I 2017 Attention is all you need *Advances in neural information processing systems* pp 5998–6008
[18] Liu J, Saxena A, Goebel K, Saha B and Wang W 2010 An adaptive recurrent neural network for remaining useful life prediction of lithium-ion batteries Tech. rep. NATIONAL AERONAUTICS AND SPACE ADMINISTRATION MOFFETT FIELD CA AMES RESEARCH ...
[19] Zhang Y, Xiong R, He H and Pecht M G 2018 *IEEE Transactions on Vehicular Technology* **67** 5695–5705
[20] Krizhevsky A, Sutskever I and Hinton G E 2012 Imagenet classification with deep convolutional neural networks *Advances in neural information processing systems* pp 1097–1105
[21] Jing L, Zhao M, Li P and Xu X 2017 *Measurement* **111** 1–10
[22] Wen L, Li X, Gao L and Zhang Y 2017 *IEEE Transactions on Industrial Electronics* **65** 5990–5998
[23] Cheng C, Ma G, Zhang Y, Sun M, Teng F, Ding H and Yuan Y 2018 *CoRR* **abs/1812.03315** (*Preprint* 1812.03315) URL http://arxiv.org/abs/1812.03315
[24] Ma G, Zhang Y, Cheng C, Zhou B, Hu P and Yuan Y 2019 *Applied Energy* **253** 113626
[25] LeCun Y, Boser B E, Denker J S, Henderson D, Howard R E, Hubbard W E and Jackel L D 1990 Handwritten digit recognition with a back-propagation network *Advances in Neural Information Processing Systems* pp 396–404
[26] LeCun Y, Boser B, Denker J S, Henderson D, Howard R E, Hubbard W and Jackel L D 1989 *Neural Computation* **1** 541–551
[27] Girshick R, Donahue J, Darrell T and Malik J 2014 Rich feature hierarchies for accurate object detection and semantic segmentation *Proceedings of the IEEE conference on computer vision and pattern recognition* pp 580–587
[28] Saxe A M, Koh P W, Chen Z, Bhand M, Suresh B and Ng A Y 2011 On random weights and unsupervised feature learning *Proceedings of the 28th International Conference on International Conference on Machine Learning* (Omnipress) pp 1089–1096
[29] Bahdanau D, Cho K and Bengio Y 2014 *arXiv preprint arXiv:1409.0473*
[30] Lai G, Chang W C, Yang Y and Liu H 2018 Modeling long-and short-term temporal patterns with deep neural networks *The 41st International ACM SIGIR Conference on Research & Development in Information Retrieval* (ACM) pp 95– 104
[31] Shih S Y, Sun F K and Lee H y 2019 *Machine Learning* **108** 1421–1441
[32] Duchi J, Hazan E and Singer Y 2011 *Journal of Machine Learning Research* 12 2121–2159
[33] Kingma D P and Ba J 2014 *arXiv preprint arXiv:1412.6980*

Acknowledgments
Authors would like to thank Prof. Ye Yuan for his valuable comments and helpful suggestions. This work was supported by the Key Laboratory of Image Processing and Intelligent Control of China under Grant IPIC2019-08 and by the National Key R&D Program of China under Grant 2018YFB1701200.

IOP Conf. Series: Materials Science and Engineering **895** (2020) 012007 doi:10.1088/1757-899X/895/1/012007

Non-contact surface roughness measurement using laser speckle technique

Dhiren R. Patel[1*], M. B. Kiran[2]

[1] Department of Mechanical Engineering, School of Technology, PDPU, Gandhinagar, 382007, Gujarat, India
[2] Department of Industrial Engineering, School of Technology, PDPU, Gandhinagar, 382007, Gujarat, India

*Email: dhirenpatel85@gmail.com

Abstract. This paper deals with selected contact type stylus method and non-contact type machine vision method using laser speckle for components prepared by grinding of AISI 1040 steel with a variety of wheels and varied depth of cut. In this interactive study, Optical method based on statistical properties of binary images is proposed for machined surfaces. Grounded metal surfaces are used to develop a binary digitized speckle pattern by a beam of He-Ne laser light source. High end camera is used to capture the image of a speckle pattern. White to black pixels ratios is computed from the binary images using image processing toolbox in Matlab. The correlation is developed between white to black pixels ratio and measured two-dimensional surface roughness parameter. Two-dimensional surface roughness parameters are recorded using a contact-type surface profilometer. The results which opted, clearly supports that these parameters have a relationship with a degree of surface roughness. A linear relationship is observed between parameter obtained from proposed technique and measured value of surface roughness using surface profilometer. The statistical analysis represents the performance of maximum relative error in prediction of surface roughness is 9%.

1. Introduction

Surface quality has been one of the most prominent measures for machined metal surfaces, in which surface roughness (Ra) is one of the crucial indices of measure. The surface's durability and manifestation viz. properties of various tribological surface, friction at the surface, wear and tear resistance against its varied applications and fatigue or rupture on the surface (notice in [1,2]), essentially depends on the properties of Surface roughness. Generally, mechanical stylus instrument is implemented in the measurement of surface roughness [3], but it induces some adverse impact by taking more duration for measurement and eventually vandalizing the surface itself [4]. Recently, non-contact measuring methods largely, optical techniques are buoyant in its application for measuring surface roughness [5]. On the other hand, these methods are gruelling due to its measuring complexity and colossal data. On bombarding the machined surface with a coherent light source, at an instance scattering occurs, causing a maze of fine defragmented spotted structure known as, speckle pattern, detect as dark and bright region [6]. This obtained pattern of dark and bright region carries crucial information regarding the surface texture. The gathered speckle pattern properties apprehends in determining the parameter of surface roughness (Ra). Non-contact type surface assessment using machine vision system includes both identification of machined texture [7] and evaluation of surface roughness [8]. Determination of surface roughness of machined surfaces by means of indirect imaging type optical methods such as interferometry, specular reflection, speckle and light scattering are reported through several studies in the literature [9–16]. S I Mohamed et al. used milled surfaces to check reliability of method of surface roughness prediction using speckle technique. In order to develop the relationship, image intensity parameters mean and standard deviation were selected. Experimental results reveled the good correlation with mean of pixel intensity while standard deviation didn't fits well [17]. Good correlation were found between surface roughness parameter and 4th level decomposed signals variance using wavelet transform of speckle images captured by complementary metal-oxide semiconductor (CMOS) camera. Signal vectors were formed using speckle images generated through illuminating laser light on the surface produced by electro discharge machining [18]. Scattering effect is prominent for coarse surface. Digitized image having white and black pixels is generated using scattering of the emerging beam reflection which forms a elicit image [15]. Experimentally, when a coarser machined surface is illuminated with laser light, occurrence of scattering effect decreases the intensity of laser beam. This variation in reflected light intensity is captured through CCD camera which is varied with change in surface roughness. The roughness measurement is based on this relationship of parameters in the procured image. The objective of the reported experimental study is to find the correlation between surface roughness parameter measured with surface profilometer and white pixel count, black pixel count and white to black pixel count ratio in order to develop the linear relationship based on correlation value which leads to prepare prediction model for average surface roughness measurement of grounded components using laser speckle technique.

2. Experimental details

2.1. Experimental set-up

Nine identical specimens were prepared using a grinding machine. A square bar made up of AISI 1040 steel was cut into nine equal sizes of 40*40 size and 17 mm length. The specimens were prepared from a square bar with 204 mm length and 40*40 mm size AISI 1040 steel which was normalized to 900 ^{0}C and hardened to 35 HRC. As shown in Table 1 different type of grinding wheel and depth of cut were selected. Later on a total, nine experimental runs have been performed with full factorial design and Handysurf 35A/B is utilized for surface roughness measurement of prepared specimens.

Table 1. Machining parameters.

Specimen No.	Grade of Wheel	Depth of Cut (mm)
1	Finish	0.1
2	Semi Finish	0.1
3	Rough	0.1
4	Finish	1.5
5	Semi Finish	1.5
6	Rough	1.5
7	Finish	2
8	Semi Finish	2
9	Rough	2

2.2. Surface roughness measurement using surface profilometer

A portable surface profilometer (Handysurf E 35A) with stylus contact method is used for measuring surface roughness parameters. In industry, contact type surface profilometer is being considered as a standard instrument in order to determine the numerous parameters of surface roughness. Measured values of amplitude parameter will be considered as point of reference to ensure the accuracy and precision of predicted measurement. Figure 1 demonstrates the setup of contact type surface profilometer. Single line scan preferred for present study. Diamond stylus is used with the radius of 5 μm at the speed of 0.6 mm/sec. Measuring conditions of surface profilometer are given in Table 2.

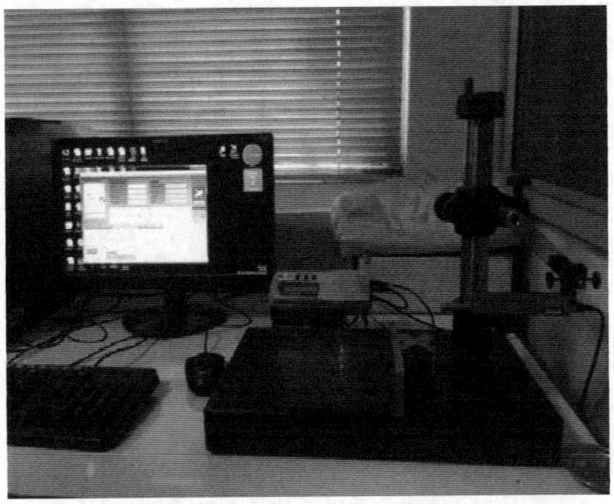

Figure 1. Surface Profilometer (Handysurf 35B).

Table 2. Measuring condition.

Measurement type	Roughness
Calculation standard	ISO'97/JIS'01/DIN
Measuring range	160 μm
Evaluation length	8 mm
Measuring speed	0.6 mm/s
Cut-off value	0.8 mm
Type of filter	Gaussian
Form remove	Straight
Material of stylus	Diamond
Radius of stylus	5 μm
Number of line scans	Single

2.3. Set-up of a machine vision system

As shown in Figure 2 machine vision system consist of the CCD camera with a high-end resolution, frame grabber, High-end industrial central processor unit is utilized for putting away and handling of the machined surfaces. The CCD camera (PULNIX) with a resolution of 1024 X 1360 is used to record scattered light. Distance between CCD camera and machined surface is kept at 130 mm. To prevent from any edge effects, 450 X 400 pixels size images were extracted from the captured images mainly from central part. The electronic noise is removed from the extracted image using noise removal algorithm written in Matlab version.

Grounded metal surface is used to develop binary digitized speckle pattern by a beam of He-Ne laser. Figure 3 shows the speckle pattern image of grounded component captured using a CCD camera. Figure 4 demonstrates that pixels in binary image with a value of 1 are displayed as white while value of 0 as black. Binary images are used for computation of parameters such as W (white pixels count) and B (black pixels count) and W/B pixels ratios.

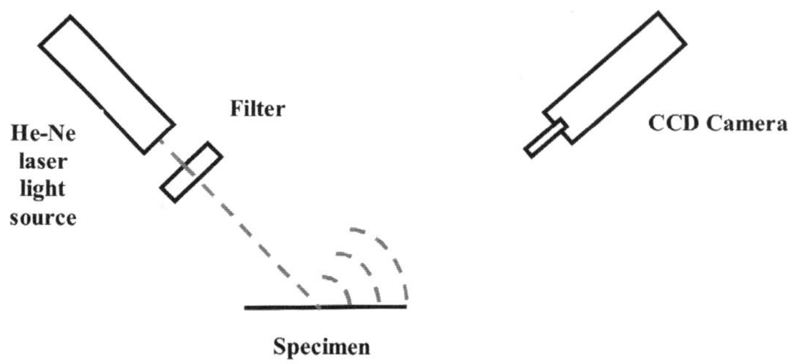

Figure 2. Setup of a machine vision system.

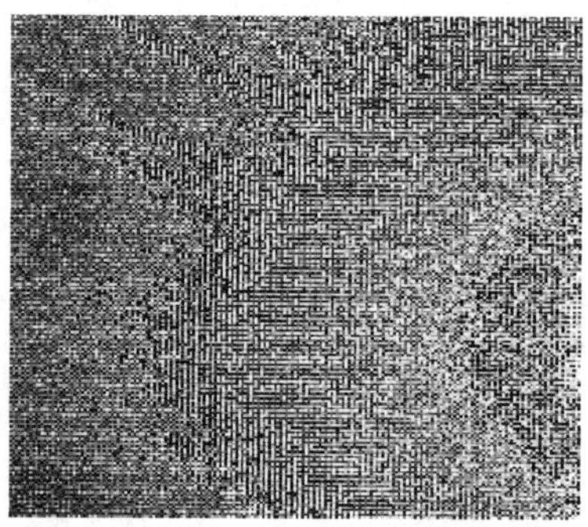

Figure 3. Speckle pattern image of grounded component.

0	0	0	1	1	1	1	0	0	0	0	0	0	0
0	1	1	1	1	1	0	1	0	0	0	0	0	0
1	0	1	1	1	0	0	1	0	0	1	0	0	1
0	0	1	1	1	1	0	0	1	0	1	1	1	1
1	0	0	1	0	1	0	0	0	0	0	0	1	1
1	1	0	0	0	0	0	0	0	0	1	0	1	1
1	1	0	0	1	0	0	0	0	0	1	0	0	1
0	1	0	0	0	0	0	0	0	0	0	1	0	1
1	0	0	0	0	0	0	0	0	0	0	1	1	1
0	0	0	0	0	0	0	0	0	0	1	1	0	0
0	0	0	0	0	0	1	0	0	0	0	0	0	0

Figure 4. Pixel values of the zoomed-in surface area show in figure 3.

3. Result and discussion

Surface roughness is measured using contact-type surface profilometer to validate the results obtained after predicting surface roughness through developing relationship between ratio of white and black pixels. The relative error between measured and predicted roughness values are shown in Table 3.

Table 3. Relative error between measured and predicted roughness value.

Measured Ra (μm)	White/Black Pixel Ratio	Predicted Ra (μm)	Absolute Error	Relative Error (%)
0.300	6.989	0.297	0.003	1%
0.310	7.591	0.337	-0.027	-9%
0.425	8.783	0.417	0.008	2%
0.515	9.780	0.484	0.031	6%
0.585	10.663	0.543	0.042	7%
0.600	11.862	0.623	-0.023	-4%
0.670	12.823	0.688	-0.018	-3%
0.715	13.612	0.741	-0.026	-4%
0.810	14.503	0.800	0.010	1%

Furthermore, a relative error has been calculated to check the measurement ability of proposed method which turns out into higher accuracy rate with maximum 9% of relative error. Experimental results show the performance level of proposed technique on grounded surface prepared under different grinding parameters, which is used to prepare good surface finished components. A good correlation has been found between measured Ra and white to black pixels ratio which is 98.97%. As shown in Figure 5 linear relationship has been developed between measured Ra (μm) and white/black pixel ratio. While developing relationship, 97.98% coefficient of determination was recorded, this provides the less relative error during prediction of surface roughness.

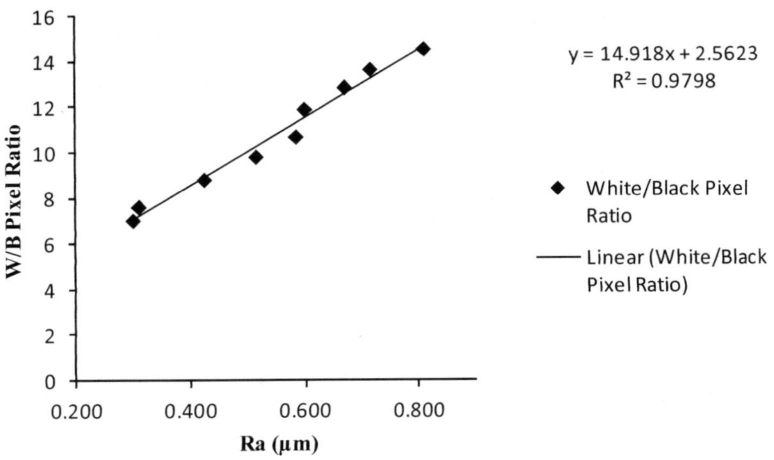

Figure 5. Linear relation between measured Ra(μm) and white/black pixel ratio.

4. Conclusion

In this study, white pixel count, black pixel count and white to black pixel ratios were computed from binary image of grounded components. Speckle pattern were created using He-Ne laser light source on machined surfaces. Linear relationship was found between image parameters and measured roughness values. The mathematical relationship is developed between white/black pixel ratio of pre-processed captured images and arithmetic mean deviation (Ra) measured through contact-type surface profilometer. The statistical analysis identifies that performance of maximum relative error in prediction of surface roughness is 9%, which showed better performance characteristics of presented linear detection model. Due to the simplicity of the method and required optical system one can conclude that it has great potential for automation and it can be utilized for in-process measurement.

5. Reference

[1] Xu D, Zhang J, Li H, Lu J, Fan Q and Dong H 2015 Research on surface topography wear of textured work roll in cold rolling *Ind. Lubr. Tribol.* **67** 269–75
[2] Xu D, Zhang J, Li H, Lu J, Wang L and Lei Z 2014 Effect of surface spacing parameters on mixed lubrication condition *Proc. Inst. Mech. Eng. Part J J. Eng. Tribol.* **228** 577–83
[3] Whitehead S A, Shearer A C, Watts D C and Wilson N H F 1999 Comparison of two stylus methods for measuring surface texture *Dent. Mater.* **15** 79–86
[4] Vorburger T V., Rhee H G, Renegar T B, Song J F and Zheng A 2007 Comparison of optical and stylus methods for measurement of surface texture *Int. J. Adv. Manuf. Technol.* **33** 110–8
[5] Al-Kindi G A and Shirinzadeh B 2007 An evaluation of surface roughness parameters measurement using vision-based data *Int. J. Mach. Tools Manuf.* **47** 697–708
[6] Kayahan E, Oktem H, Hacizade F, Nasibov H and Gundogdu O 2010 Measurement of surface roughness of metals using binary speckle image analysis *Tribol. Int.* **43** 307–11
[7] Patel D R, Vakharia V and Kiran M B 2019 Texture classification of machined surfaces using image processing and machine learning techniques *FME Trans.* **47** 865–72
[8] Patel D R and Kiran M B 2019 Experimental analysis on relationship between roughness parameters and texture features of 6082T6 sandblasted components *Materials Today: Proceedings* vol 18(Elsevier Ltd.)pp 3008–16
[9] Kumar M, Sharma A, Shahi A S, Pwta Á S Á and Butt Á 2019 *Surface Roughness Evaluation of Milled Surfaces by Image Processing of Speckle and White-Light Images* (Springer Singapore)
[10] Mahashar Ali J and Murugan M 2017 Surface roughness characterisation of turned surfaces using image processing *Int. J. Mach. Mach. Mater.* **19** 394–406
[11] Dias M R B, Dornelas D, Balthazar W F, Huguenin J A O and da Silva L 2017 Lacunarity study of speckle patterns produced by rough surfaces *Phys. A Stat. Mech. its Appl.* **486** 328–36
[12] Dev K, A. S. G P, H A, P P, Chan K H K and Matham M V. 2017 Surface roughness measurement of additive manufactured samples using angular speckle correlation *Fifth Int. Conf. Opt. Photonics Eng.* **10449**104492W
[13] Bharathi S and Ratnam M M 2019 Evaluation of 3D Surface Roughness of Milled Surfaces using Laser Speckle Pattern *IOP Conf. Ser. Mater. Sci. Eng.* **530**
[14] Balamurugan R and Rajarajan G 2017 Study of drying process of paint by dynamic speckle with B/D pixel counting technique *Opt. Lasers Eng.* **98** 62–8
[15] Xu D, Yang Q, Dong F and Krishnaswamy S 2018 Evaluation of surface roughness of a machined metal surface based on laser speckle pattern *J. Eng.* **2018** 773–8
[16] Rodrlguez F, Cotto I, Dasilva S, Rey P and der Straeten K Van 2017 Speckle characterization of surface roughness obtained by laser texturing *Procedia Manuf.* **13** 519–25
[17] Mohamed Suhail S I, Mahashar Ali J, Siddhi Jailani H and Murugan M 2018 Vision based system for surface roughness characterisation of milled surfaces using speckle line images *IOP* Conf. Ser. Mater. Sci. Eng. 402
[18] Mahashar Ali J, Siddhi Jailani H and Murugan M 2020 Surface roughness evaluation of electrical discharge machined surfaces using wavelet transform of speckle line images *Meas. J. Int. Meas. Confed.* **149** 107029

IOP Conf. Series: Materials Science and Engineering **895** (2020) 012008 doi:10.1088/1757-899X/895/1/012008

Multipurpose Agricultural Robot Platform

Sunti Thaitanun[1], Sitthivich Loetthanaphan[1], Sangdad Wanichanan[1], Ye Moe Myint[2,3], Gridsada Phanomchoeng[1,4,*], Ratchatin Chancharoen[1,3], Lunchakorn Wuttisittikulkij[1,3], Muhammad Saadi[5], and Sanika K. Wijayasekara[6]

[1]Department of Mechanical Engineering, Chulalongkorn University, Thailand
[2]Department of Electrical Engineering, Chulalongkorn University, Thailand
[3]Smart Wireless Communication Ecosystem Research Group, Chulalongkorn University
[4]Applied Medical Virology research unit, Department of Microbiology, Chulalongkorn University.
[5]Department of Electrical Engineering, University of Central Punjab, Pakistan
[6]Department of Electrical Electronic and Telecommunication Engineering, General Sir John Kotelawala Defence University, Ratmalana, Sri Lanka.

*Corresponding author: gridsada.phanomchoeng@gmail.com

Abstract. During the last decade, agricultural sector has witnessed a wide adoption of cutting edge technology not only to improve the quality and quantity of the crops but also to ease the labour burden. These technological oriented and innovative agricultural solutions have opened new horizons for opportunists, researchers and investors along with the creation and development of new markets. Robotic and automation technologies are now complementing farmers in various ways. In this paper, we present the mechanical and electrical design of multi-purpose agricultural robot which has the capabilities to move in three Cartesian coordinate axis along with the functionality of harvesting small sized fruits with the help of grippers. Our developed robot prototype can be controlled via remote control. Prototype testing results show that the developed robot can reach to the target by passing any small sized obstacle and harvest the fruit. We believe that such low cost, less complex and indigenously developed multipurpose agricultural robots can help farmers by reducing their labour cost and human effort.

1. Introduction

The agriculture industry provides the world with one of the most basic needs: Food. "End hunger, achieve food security and improved nutrition and promote sustainable agriculture" is the second sustainable development goal of the United Nations (UN) by 2030 [1]. Countries like India, Thailand, Pakistan etc., where a vast percentage of the work force is associated with agriculture-related sectors contributes less than 10% of the gross domestic product (GDP) to the country's economy [2]. In Thailand, more than 28% of the total work force is associated with agriculture either directly or indirectly. Interestingly, the dominant portion of the agricultural labour force aged more than 40 years with less youth adopting this hard profession [3]. The off-balance problem between labours and agricultural product values and the rise in the proportion of the older workers is impacting production negatively. The labour shortage along with the increase of world population essentially needs technological advancements to meet growing and shifting demands. There has been a dramatic shift in the use of technology in the agriculture industry in the last decade, however, farmers from the developing countries are still lacking in modern tools usage primarily because of cost, unavailability of products in local market and inadequacy of education. Human force is generally used for all the steps of crop growth starting from sowing, irrigation, monitoring the growth, pest control and harvesting. The demand for food is outpacing available farmland and robotic application such as agricultural robot is one of the ways that helps farmers to close this gap. The agricultural robots are increasing production yields for farmers in various ways where the drones are used to analyse the condition of farming environment and autonomous tractors for sowing, mowing, fertilizing, etc. and robotic arms to planting, harvesting and picking, the technology is being deployed in creative and innovative applications. Agricultural robots carry out slow, repetitive and dull tasks for the farmers, allowing them to focus more on improving overall agriculture production yields. Some of the most common robots in agriculture are used for harvesting and picking, weed control, autonomous mowing, pruning, seeding, spraying and thinning, phenotyping, sorting and packing. Among these, harvesting and picking is one of the most popular robotic applications in agriculture due to the accuracy and speed of the robots can achieve to improve the size of yields and reduce waste from plants being left in the field.

In this work, the aim is to build a strong mobile prototype structure of an agricultural robot. The robot will perform harvesting and picking small fruits by the help of the driving system that is designed to support the robot to be stable on the different kinds of soil surface condition. The robot will also be used for many other agricultural functions in the future by installing the desired additional applications such as watering, sowing seeds, spraying pesticides in the system.

2. Overall System Design

Overall system design is constructed mainly by mechanical system design and electrical system design as shown in Figure 1. The mechanical part is responsible for the movement of robot in three axis which is done with the help of motors and wheels. In addition to the movement, mechanical system is also responsible for governing the harvesting system i.e. plucking the fruit from the plant. The electrical system works in collaboration with the mechanical system. The electrical system is responsible for switching ON and OFF, speed up and down the motors which are installed in the robot which facilitates the robot to move in X, Y and Z directions. The Electrical system is also responsible for establishing a communication channel between the remote control and the Arduino. The details of each system design will be discussed in the following sections.

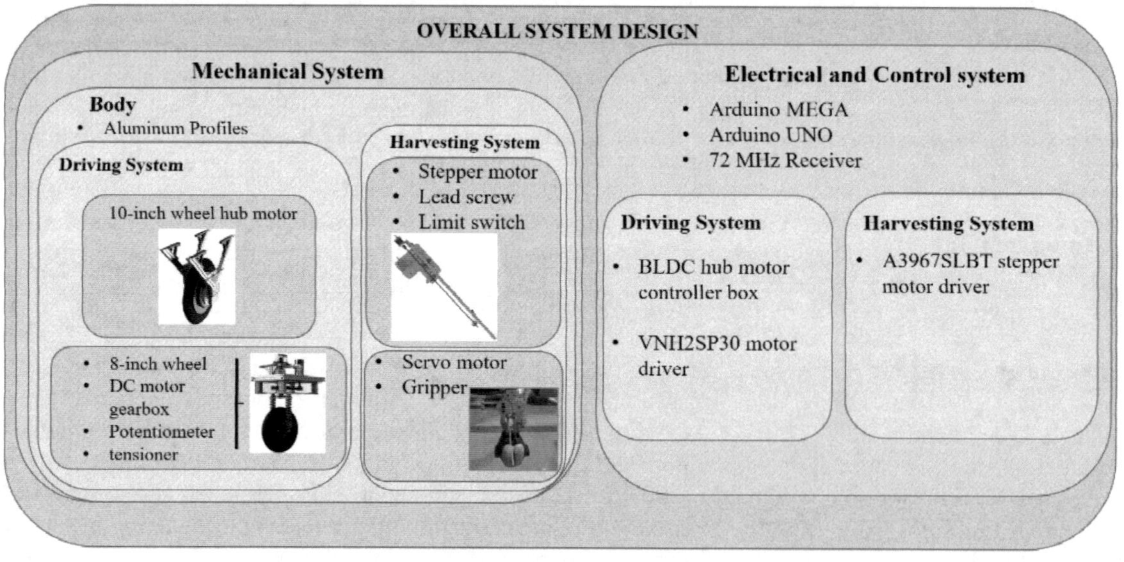

Figure 1. Overall System Design of the Robot.

The platform structure, which is shown in Figure 2, has a size of 120 x 105 x 96 cm and weights 60 kg to fit the fruit plot and is high enough to attach accessories without colliding with the plants.

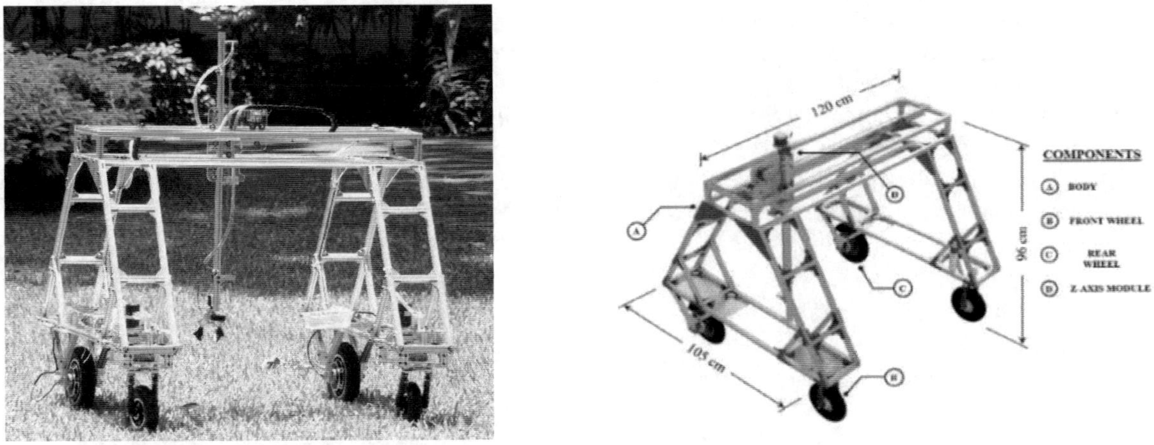

Figure 2. Overall Platform Structure of Robot.

3. Mechanical System

In the design of robot structure, aluminium profile is chosen as it can be easily constructed for desired structures due to its flexibility, durability and light weight feature. For mobility, the robot has four wheels in total: two rear wheels each having 8 inches diameter along with DC motors for drive system and two 10 inch diameter front wheels with Brushless DC (BLDC) Hub motors for turning. There is a gripper to harvest fruits at the middle of robot that can be moved with the help of Y-axis and Z-axis modules.

3.1. Front Wheels

The front wheels are used for turning the robot with the help of motors connected to the belt. The turning degree of wheel is important to consider because the inside wheels must have more turning degrees that the outer wheels. If both wheels turn at the same angle, the inner wheels will cause stings and trying to push the robot head out of the turning circle causing the wheel to slip and the turn is not smooth.

From the calculations in Figure 3, it was found that when making the inner wheel to have the maximum turning degree at 50 degrees, the outer wheel must have a turning angle of about 27 degrees to allow the wheel to rotate smoothly and not to slip.

3.2. Rear Wheels

The rear wheel are used to move the robot forward and backward by controlling BLDC Hub motor. Aluminium profiles support to hold the wheel next to the frame of the robot handling the weight of the robot on both sides.

3.3. Z-Axis Module

For Z axis, the "Lead Screw" mechanism is used for up and down movements which needs to be a lot more precise than Pinon belt. A lead screw has a thread optimized for reducing friction and is used frequently in many applications. Precise positioning and non-overhauling can be obtained with a shorter lead, however, the main downside of lead screws is that they are mechanical inefficiency converting 20% to 80% of the applied torque into linear thrust and rest of the energy is wasted in the form of heat [4]. Furthermore, the lead screw mechanism is made self-locking as well which will prevent any accident or damage to the fruits.

3.4. Gripper

For the purpose of harvesting fruits, we used the gripper to pull an orange fruit out of the plant which can be done by the handle attached with servo motors and this can move on 2 axes in the plane, -90 degrees to 90 degrees. The clamping area can be opened to clamp the orange fruit using force to pull the orange fruit not more than 10 Newton. It will cause the orange fruit to fall out of the plant and no bruises on the orange fruit. Since two main parallel surfaces of contact are used, the fruit suffers from high concentrated compressive force. In order to distribute this force, taking into account the variation in size and shape of the strawberry the contact surfaces must be increased [4]. This solution is obtained through the simulation that when each side of the handle receives different forces which will cause the handle to damage the fruit or not. The designed gripper for the platform is shown in Figure 4.

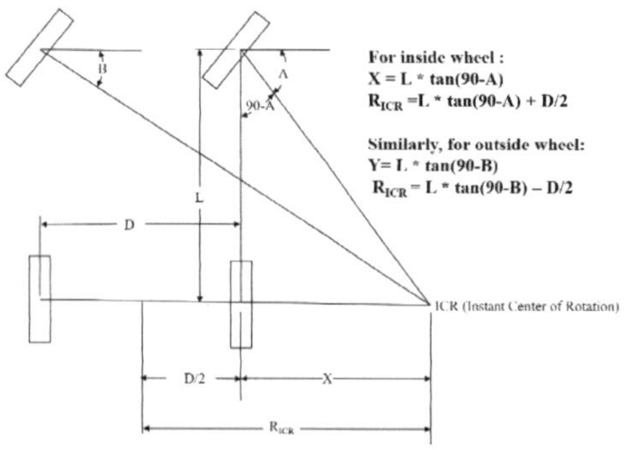

Figure 3. Principles of calculating the degree of turning wheel.

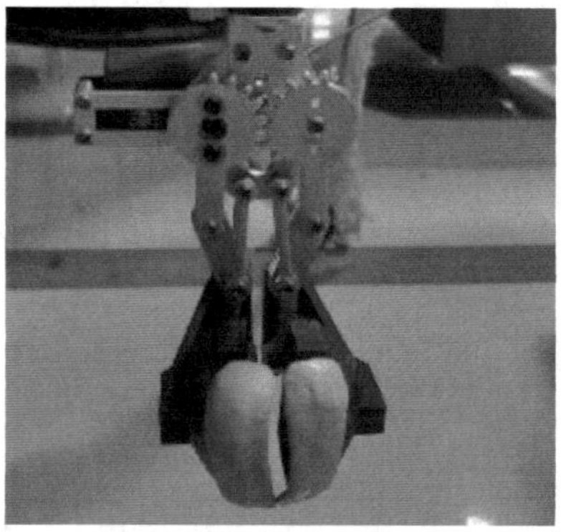

Figure 4. Gripper.

4. Electrical and Control System

Components of electronic circuits used to control the robot are shown in Figure 5. For controlling the robot, it will use the manual system controlled by the Futaba brand remote control with all 6 channel values which can be divided as follows: Channels 1 and 4 can move freely in the horizontal direction. Channels 2 and 3 can move freely vertically. Channels 5 and 6 can move only 2 strokes which is up and down. System controls the front wheels consists of 2 DC Motors, 2 Motor Drive VNH2SP30 and 2 potentiometers. When Arduino receives a signal from the receiver, Channel 1 will take control of motor rotation direction attached to the front wheel. Based on Figure 3, knowing that both wheels must rotate at unequal speeds. For such conditions, the installation of a potentiometer into the front wheel turning mechanism makes it possible to know where the wheels are. After that, the Arduino board receives signals from the receiver and sends the signal to the VNH2SP30 board to control the speed of each motor. When the wheel spins continuously, the potentiometer will dictate to which extent each wheel can rotate. The control system for Y-axis and Z-axis modules consists of two NEMA17 stepper motors, two A3967SLB Stepper Motor Driver T and 4 limit switches. The motor is used as a bipolar type stepper motor so it requires a dedicated board to control, when the board runs at one loop, the motor will rotate about one pulse. In this case, the NEMA17 stepper motors will rotate about 1.8 degrees per 1 pulse. Therefore, the Arduino board is used to receive signals from Channel 3 for the Z-axis and Channel 4 for the Y-axis. After that, the values will be used for output mapping of the signal to the motor to control the motor speed for the intended operation. In addition, there are 2 limit switches for the Y-axis and two for the Z-axis to determine the movement of the motor and prevent the damage that may occur. Control system for the rear wheels contains 2 controller boxes, 2 BLDC Hub Motor. The rear wheel, which acts as driving wheel, has a brushless DC motor. The handle consists of 2 Servo Motor, MG955 that determine the position of the handle can be controlled by ordering directly through the Arduino board. Arduino takes the value obtained from mapping process and then orders a servo motor via digital channel to get desired degree rotation. In the second motor, only the value received from Channel 2 will be used to control the position of the motor at 130 degrees (the handle will open) with 180 degrees (the handle will close) only. A 72 MHz receiver is used to send the signal from the remote control to Arduino board and connect the receiver's output into the digital into of the Arduino board.

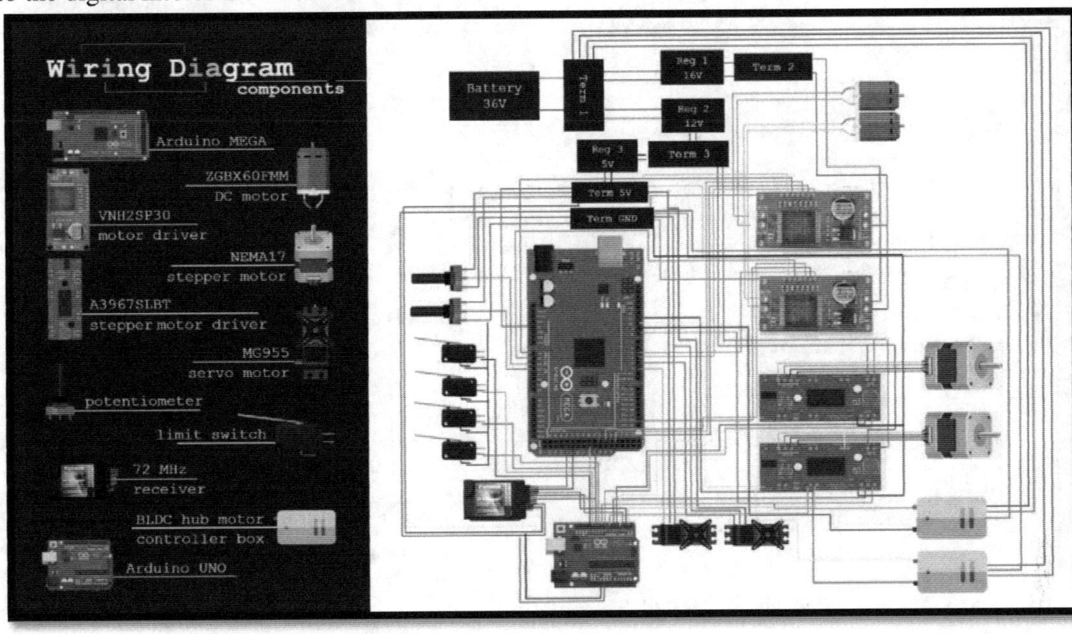

Figure 5. Wiring Diagram for Electrical and Control System

5. Prototype Testing

In basic test, the basic abilities of the robot are tested one by one. The test is divided into 5 main parts: (1). Testing the robot if it can run in a straight line. To do so, both rear wheels must rotate at the same speed and force the front wheels to be parallel to facilitate robot run. The distance of this test was approximately 5 meters. (2). Testing the robot if it can turn: it was done to verify that the principle and the results obtained from the calculation in Figure 5 can be used and check that both wheels at the front can be moved without any skidding. Along with finding the minimum radius of curvature that is estimated so that the robot can do when the inner corner of the wheels at the turn that has a value of 50 degrees. (3). Testing if the robot can cross obstacles: this test is done to verify that the robot has enough power to run across small obstacles and test the impact reduction of the vibration reduction system installed on both front wheels. (4). Testing if robot can park or stop in the designated area: it was done to check the driving system of robot that can move in a desired location. In this test, the robot moves for a distance of about 6 meters and can be stopped in the specified area. (5). Testing Y-axis and Z-axis modules of the robot: the test was done to verify that the two control motors can move smoothly and find the working area of the Y-axis and the Z-axis module using the limit switch to determine the extent of all movement. Overall testing which is a test to see the overall ability of the robot form a stationary position to do harvesting in a specified time. The score criteria for overall testing is shown in Table 1. All the example of the test is shown in Figure 6.

Table 1. Scoring criteria for overall testing.

Criteria used	**Score**	**Remark**
Able to collect fruits	2	
Damage	-1	With a stick or bruise, etc.

(a)

Performance	Score	Remark
Excellent	10+ pts	
Good	7 to 9 pts	
Fair	4 to 6 pts	
Poor	0 to 3 pts	

(b)

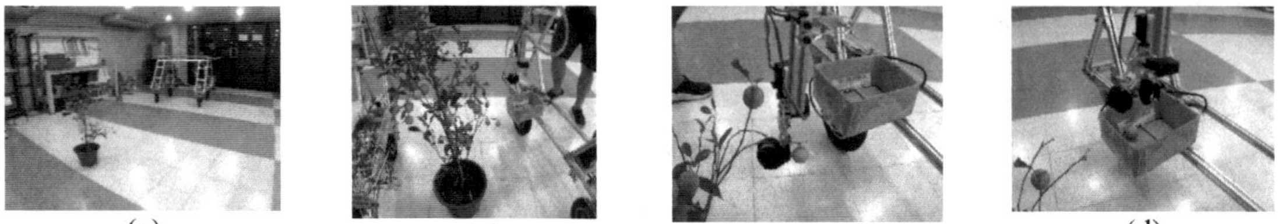

(a) (b) (c) (d)

Figure 6. Overall testing (a) Robot is at the starting point (b) Robot is at the targeted plant (c) Harvesting (d) The fruit is stored in the box.

6. Discussion and Testing Results

Based on various tests in Section 5, the results of the tests are shown in Table 2. Total score is 8 points from the criteria considering that the test result was good.

Table 2. Results from Tests.

Testing	Result	Remark
Basic Testing		
Can run in a straight line at the specified path	Pass	1.5 meters length
Can turn	Pass	The radius of curvature is about 1.4 meters. (Measured from the middle of the robot)
Can run across obstacles that are simulated to compare with rough areas	Pass	acrylic sheets about 2 cm high are used as obstacles
Can park or stop in the designated area	Pass	The area is approximately 1.2x1.5 meters.
Can move the Y axis and the Z axis module without interruption	Pass	
Overall Testing		
Test all parts at the same time and collect fruits	8 points	Can collect 4 fruits without damaging them

7. Conclusion

In this paper, we designed a multipurpose agricultural robot. The indigenously developed has two main parts and they are the mechanical system and the electrical system. The robot platform structure mainly uses aluminium profile material due to strength, durability, corrosion resistance, flexibility and lightness. This research work presents a preliminary study on the part of harvesting fruits using Orange as an example. The developed prototype was rigorously tested and form the test results, it can be found out that the robot can perform basic tasks like moving forward, backward, turning left and turning right according to the intended purpose using two rear wheels for driving system and two front wheels for turning the root. Also with the help of Z-axis module, it can harvest fruits which are not very large by size. The platform of this versatile robot for agriculture is designed with a purpose to be a model for further development in the future, such as putting more various agricultural functions and also the introduction of image processing technology to enable robots to work automatically with precise positioning of the robot etc.

8. References

[1] Jones A D and Ejeta G 2016 A new global agenda for nutrition and health: the importance of agriculture and food systems Bulletin of the World Health Organization, 94(3) p 228.
[2] Office of the National Economic and Social Development Council 2019 Gross domestic product : Q4/2018 Bangkok
[3] National Statistical Office 2017 Summary of the labor force survey in Thailand : March 2017 Bangkok
[4] Dimeas F, Sako D V, Moulianitis VC and Aspragathos N A 2013 Towards designing a robot gripper for efficient strawberry harvesting 22nd International workshop on robotics in Alpe-Adria-Danube region (RAAD 2013).

Acknowledgments
This work was a part of the Intelligent 3D Printing for Product Development Project which was supported by the 2nd Century Development Fund, Chulalongkorn University.

IOP Conf. Series: Materials Science and Engineering **895** (2020) 012009 doi:10.1088/1757-899X/895/1/012009

Static characteristics of a simple rotor-bearing system: modelling and experiment

Phuoc Vinh Dang[1], Le Dinh Minh Nhat[2], Nhu Thanh Vo[1], Thanh Nghi Ngo[1], Hoai Nam Le[1], Huu Nhan Nguyen[1]

[1]Department of Mechanical Engineering, The University of Danang – Danang University of Science and Technology.
[2]School of Material Science - Japan Advanced Institute of Science and Technology.

Email: dpvinh@dut.udn.vn

Abstract. The dynamics of rotating machinery have been extensively investigated during the past century. To date, longer rotating shafts, higher rotating velocity, higher fluid pressure, and generally high performances are required. For this purpose, a deep understanding and a more accurate modelling of the rotating machinery dynamical behaviour is needed, especially in terms of the resultant vibrations and stability issues. In this paper, a mathematical model was developed for a simple rigid rotor-bearing system using finite element method. Several natural frequencies and corresponding mode shapes of the systems were presented. In order to validate the model, some experimental tests were performed using a small test rig. The response of the system in the frequency domain and the direct orbit of the shaft were plotted as well.

1. Introduction

The dynamics of rotating machinery have been extensively investigated during the past century. To date, longer rotating shafts, higher rotating velocity, higher fluid pressure, and generally high performances are required. For this purpose, a deep understanding and a more accurate modelling of the rotating machinery dynamical behaviour is needed, especially in terms of the resultant vibrations and stability issues.

During the operation of a rotational mechanical system supported on hydrodynamic bearings, hydrodynamic pressure distribution is generated within the bearing oil film, which bears both weight and unbalanced forces of the rotor and has a direct influence on the dynamic behaviour of the rotor i.e., system stability, bending critical speed, vibration modes.

Many literatures can be found about identification methods of rotor-bearing dynamic parameters [1]-[4]. R. Tiwari [5] reported the development of algorithms for the simultaneous identification of both the unbalance and the bearing dynamic parameters; the proposed algorithms were compared against experimental data. Ref. [6] proposed a technique known as algebraic identification, which allows the development of identifiers aimed to determination on-line of unknown parameters, based on both the differential algebra and the operational calculus from the system mathematical model. Refs [7] proposed identifiers using algebraic identification to estimate the rotor dynamic system parameters; both numerical and experimental results showed that the identification was obtained in a short time and exhibits high robustness before parametric uncertainty. One of the main advantages of the algebraic identification is to provide identification relationships completely independent from the

initial conditions of the system since only requires the system response signal. The estimation of the parameters is carried out online in either continuous or discrete time.
In this paper, a mathematical model was developed for a simple rigid rotor-bearing system using finite element method. Several natural frequencies and corresponding mode shapes of the systems were presented. In order to validate the model, some experimental tests were performed using a small test rig. The response of the system in the frequency domain and the direct orbit of the shaft were plotted as well.

2. The mathematical model

It is well-known that the dynamic behaviour of rotating machines is strongly influenced by the bearings properties since the stiffness of the rotor-bearing system is the result of the bearing stiffness and the shaft stiffness addition. Likewise, the damping of the system is mainly the result of the bearing damping properties. A simple rigid rotor-bearing system is shown in **Figure 1**. The rotor is supported on two different flexible bearings.

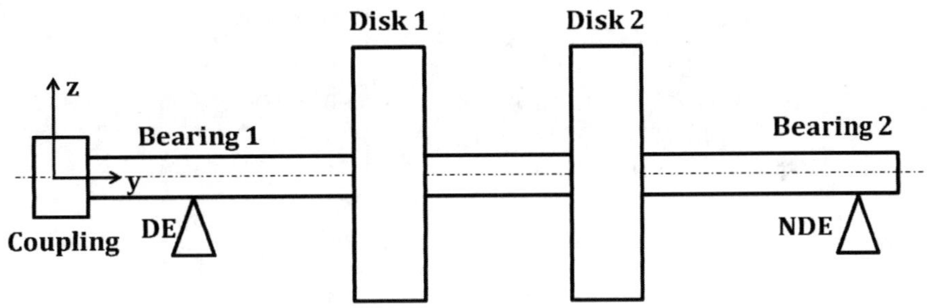

Figure 1. Rotor-bearing system.

The finite element method was used for the numerical simulation with 4 degree of freedoms (DOFs) for one node i.e., radial displacement (u,w) and angular displacement (θ,ψ) in y and z direction (see **Figure 2**). Considering *j-th* element, the displacement respond vector as follow:

$$\mathbf{x}_j = \left\{ u_j \quad w_j \quad \theta_{x_j} \quad \psi_{z_j} \quad u_{j+1} \quad w_{j+1} \quad \theta_{x_{j+1}} \quad \psi_{z_{j+1}} \right\} \tag{1}$$

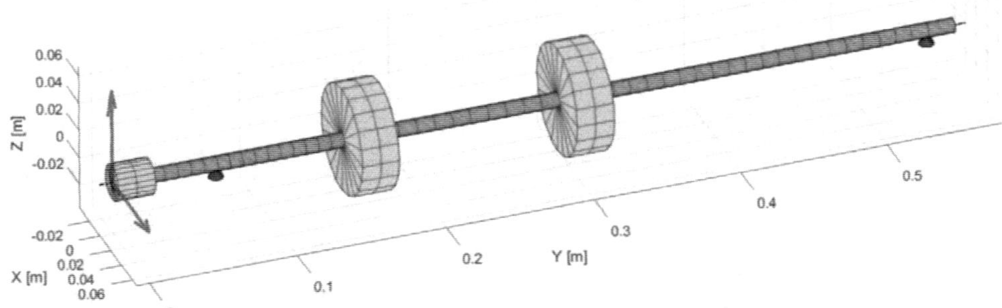

Figure 2. Numerical model for rotor-bearing system.

The mathematical model corresponding to the two DOFs rotor-bearing system showed in, considering constant frequency is given by:

$$[\mathbf{M}]\ddot{\mathbf{x}} + ([\mathbf{C}] + \Omega[\mathbf{G}])\dot{\mathbf{x}} + [\mathbf{K}]\mathbf{x} = W + F(t) \tag{2}$$

where Ω is the shaft rotational speed; W is rotor weight and F is external force; [**M**] and [$\mathbf{M_s}$] are mass and secondary effect of rotatory inertia matrix; [**C**] is damping matrix; [**G**] is gyroscopic matrix and [**K**] is stiffness matrix, with:

$$[\mathbf{M}]_{8\times8}=\frac{\rho S_m L}{420}\begin{bmatrix} 156 & 0 & 0 & -22L & 54 & 0 & 0 & 13L \\ 0 & 156 & 22L & 0 & 0 & 54 & -13L & 0 \\ 0 & 22L & 4L^2 & 0 & 0 & 13L & -3L^2 & 0 \\ -22L & 0 & 0 & 4L^2 & -13L & 0 & 0 & -3L^2 \\ 54 & 0 & 0 & -13L & 156 & 0 & 0 & 22L \\ 0 & 54 & 13L & 0 & 0 & 156 & -22L & 0 \\ 0 & -13L & -3L^2 & 0 & 0 & -22L & 4L^2 & 0 \\ 13L & 0 & 0 & -3L^2 & 22L & 0 & 0 & 4L^2 \end{bmatrix} \quad (3)$$

$$[\mathbf{G}]_{8\times8}=\frac{\rho I}{15L}\begin{bmatrix} 0 & -36 & -3L & 0 & 0 & 36 & -3L & 0 \\ 36 & 0 & 0 & -3L & -36 & 0 & 0 & -3L \\ 3L & 0 & 0 & -4L^2 & -3L & 0 & 0 & L^2 \\ 0 & 3L & 4L^2 & 0 & 0 & -3L & -L^2 & 0 \\ 0 & 36 & 3L & 0 & 0 & -36 & 3L & 0 \\ -36 & 0 & 0 & 3L & 36 & 0 & 0 & 3L \\ 3L & 0 & 0 & L^2 & -3L & 0 & 0 & -4L^2 \\ 0 & 3L & -L^2 & 0 & 0 & -3L & 4L^2 & 0 \end{bmatrix} \quad (4)$$

$$[\mathbf{K}]_{8\times8}=\frac{EI_k}{(1+a)L^3}\begin{bmatrix} 12 & 0 & 0 & -6L & -12 & 0 & 0 & -6L \\ 0 & 12 & 6L & 0 & 0 & -12 & 6L & 0 \\ 0 & 6L & (4+a)L^2 & 0 & 0 & -6L & (2-a)L^2 & 0 \\ -6L & 0 & 0 & (4+a)L^2 & 6L & 0 & 0 & (2-a)L^2 \\ -12 & 0 & 0 & 6L & 12 & 0 & 0 & 6L \\ 0 & -12 & -6L & 0 & 0 & 12 & -6L & 0 \\ 0 & 6L & (2-a)L^2 & 0 & 0 & -6L & (4+a)L^2 & 0 \\ -6L & 0 & 0 & (2-a)L^2 & 6L & 0 & 0 & (4+a)L^2 \end{bmatrix} \quad (5)$$

where ρ, S_m, L, E, I_k and a are the density, cross-section, length, Young's modulus, moment of inertia of the cross section and shear coefficient of the *j-th* element.

3. Experiment setup

Figure 3 presents an overview of the experimental test-rig using for the experimental tests. Basically, it is encompassed by a flexible rotor-bearing system.

In this test rig, a rigid shaft of 10 mm diameter which is supported by two rolling element bearings at the ends of shaft is driven by an electric motor through a flexible coupling. The coupling is modified from a normal flexible cross coupling in order to minimize the coupling constraints on the rotor lateral movement. The DC motor can run at a maximum speed of 5000 rpm. One or two disks with a diameter and thickness of 75 mm and 25 mm respectively can be installed on the shaft. Two proximity probes of 90° phase angle are installed at each bearing position in the horizontal, named as P1X and P2X, and vertical direction, named as P1Y and P2Y, to measure the vibration of the shaft (see **Figure 4**). One additional proximity probe is used as a key phasor for tacho signal and synchronous averaging technique [9]-[10].

In this study, the shaft was run-up gradually from 250 rpm up to 4000 rpm first. Next, it was remained at the 4000 rpm about 20 second after run-down slowly to 250 rpm. During the test, signal from 4 proximity probes were acquired with the sampling frequency of 10 kHz.

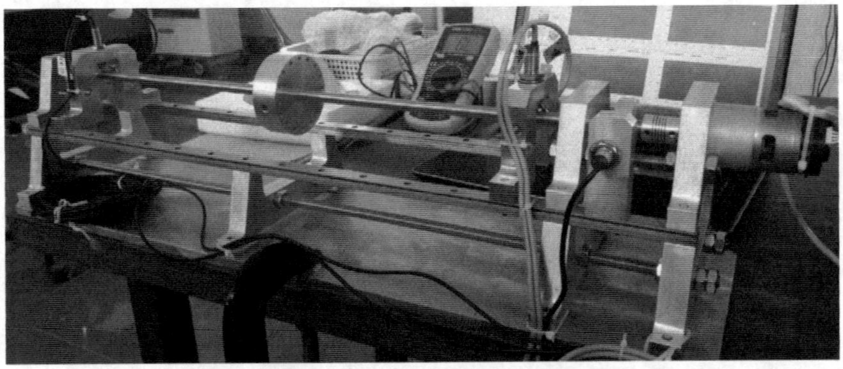

Figure 3. Photo of the test rig using for the experiment.

Figure 4. Installed proximity sensors for vibration measurement [8].

4. Results

In the present section, the natural frequencies and mode shapes of the system obtained from the numerical simulation have been presented. In order to validate the model, some experimental tests were performed using a small test rig which were described in the previous section. With no running and neglecting the weight of rotor, Eq. (2) can be expressed as:

$$[\mathbf{M}]\ddot{x} + [\mathbf{K}]x = 0 \tag{6}$$

By re-written Eq. (6), we have:

$$-[\mathbf{M}]^{-1}[\mathbf{K}]x = \omega^2 x \tag{7}$$

Then, the natural frequencies can be obtained by Eq. (7)

$$\omega = \sqrt{[\mathbf{M}]^{-1}[\mathbf{K}]} \tag{8}$$

First-six natural frequencies of the system are listed in Table 1.

Table 1. First-six natural frequencies of the tested system.

Mode	#1	#2	#3	#4	#5	#6
Natural frequency (Hz)	25,8078	25,8458	110,4473	110,7843	347,1279	357,4829

IOP Conf. Series: Materials Science and Engineering **895** (2020) 012009 doi:10.1088/1757-899X/895/1/012009

Figure 5 shows the 3D mode shape of the rotor-bearing system. Because of the sake of simplicity, only mode shapes of the system at the first and third natural frequency were plotted. It can be seen that the maximum displacement of the shaft at mode #1 and mode #3 is approximately 200 μm and 100 μm, respectively.

Figure 6 shows the experimental signals of two proximity sensors in the frequency domain when the shaft was run-up from 250 rpm to 4000 rpm. This plot is obtained by performing the Fast Fourier Transform (FFT) technique to change to the signal from the time domain to the frequency domain. It is possible to see from the **Figure 6** that the first and second nature frequency of the system are approximately 23 Hz and 108 Hz. These results show a good agreement with the simulation values. This identification is very important to avoid the resonance of the system. In order to obtain exactly the natural frequency of the system, the impact test is usually performed.

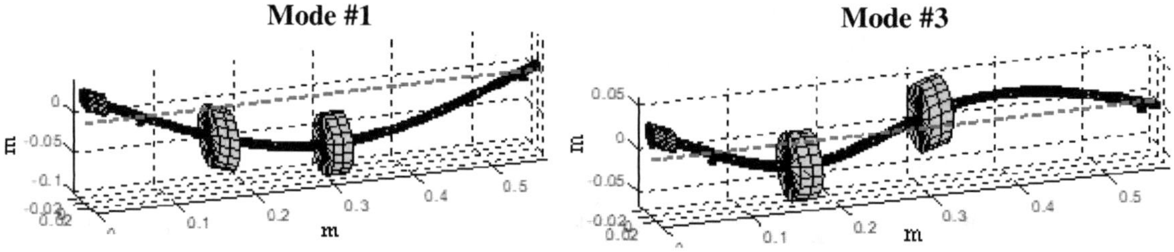

Figure 5. Mode shapes obtained from numerical simulation.

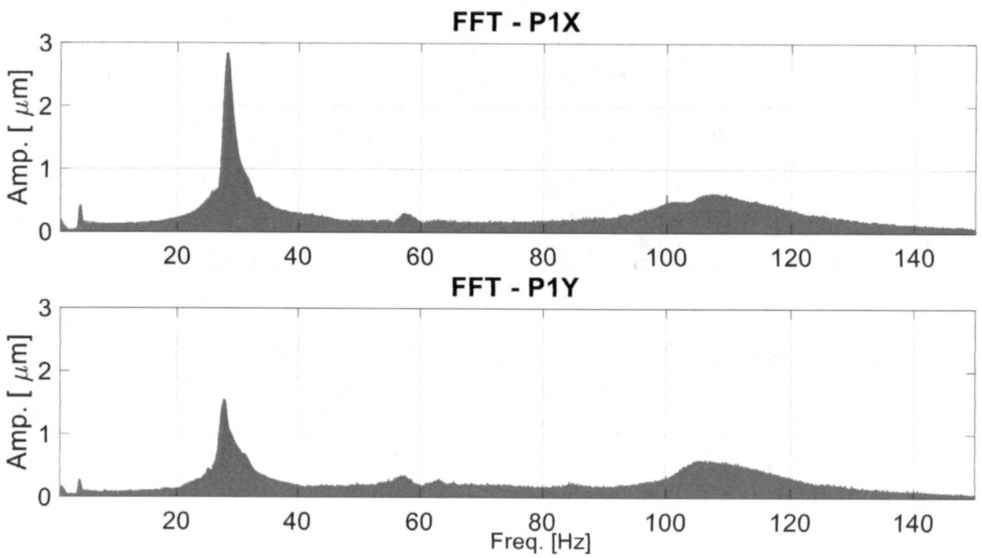

Figure 6. Experimental signal from P1X and P1Y in the frequency domain.

Figure 7 shows the orbit the shaft plotted from proximity probes in the non-driven end (NDE) bearing and driven end (DE) bearing at the constant speed of 4000 rpm. It is clear that the orbit at the DE bearing has irregular shape. This fact is probably due to the flexibility of the system. At the NDE position, the orbit has a quite smooth circle shape.

Figure 8 shows the experimental 1X orbit of the shaft at two bearing positions with an elliptical form. These orbits correspond to the vibration of shaft at a frequency of 66.7 Hz (about 4000 rpm). It should be noted that the vibration amplitude at the DE position is nearly two times compared to the NDE position.

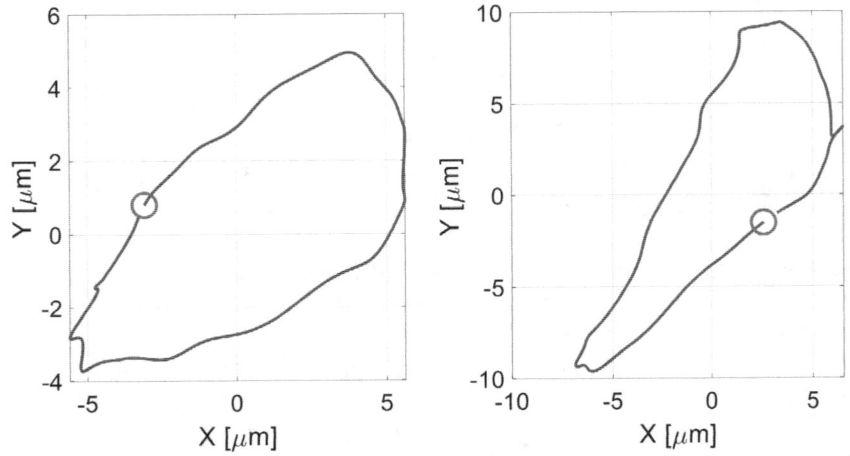

Figure 7. Experimental direct orbit of shaft at the NDE (left) and DE (right) position.

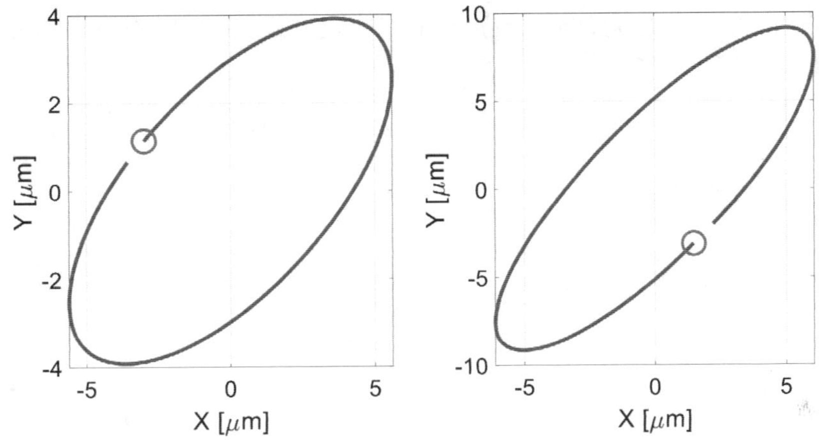

Figure 8. Experimental 1X orbit of shaft at the NDE (left) and DE (right) position.

5. Conclusions

This paper developed a mathematical model for a simple rigid rotor-bearing system using finite element method. Several natural frequencies and corresponding mode shapes of the systems were presented. In order to validate the model, some experimental tests were performed using a small suitable test rig. The response of the system in the frequency domain and the direct orbit and 1X orbit of the shaft were plotted as well.

6. References

[1] Chatterton S, Pennacchi P, Dang P.V and Vania A 2014. A test rig for evaluating tilting-pad journal bearing characteristics. *Mechanisms and Machine Science* **21,** 921-930.

[2] Dang P.V, Chatterton S, Pennacchi P and Vania A 2016. Effect of the load direction on non-nominal five-pad tilting-pad journal bearings. *Tribology International*. **98**, 197-211.

[3] Dang P.V, Chatterton S, Pennacchi P and Vania A 2018. Numerical investigation of the effect of manufacturing errors in pads on the behaviour of tilting-pad journal bearings. *Proceedings of the Institution of Mechanical Engineers, Part J: Journal of Engineering Tribology*. **232**, 480-500

[4] Chatterton S, Pennacchi P, Vania A, Luca A.D and Dang P.V 2019. Tribo-design of lubricants for power loss reduction in the oil-film bearings of a process industry machine: Modelling and experimental tests. *Tribology International* **130,** 133-145.

[5] Tiwari R, Chakravarthy V 2009. Simultaneous Estimation of the Residual Unbalance and Bearing Dynamic Parameters from the Experimental Data in a Rotor-Bearing System. *Mechanism and Machine Theory* **44**, 792-812.

[6] Fliess M, Hebertt R 2003. An Algebraic Framework for Linear Identification. *ESAIM: Control, Optimisation and Calculus of Variations* **9**, 151-168.

[7] Manuel A, Carvajal B, Navarro S 2014. On- Line Algebric Identification of Eccentricity Parameters in Active Rotor-Bearing Systems. *International Journal of Mechanical Sciences* **85**, 152-159.

[8] Pham N.Q.H, Dang P.V, Ngo T.N, Tran P.T 2019. Design of a small-scale test rig for rotating machinery characterization. *Lecture Notes in Networks and Systems* **63**, 229-235.

[9] Chatterton S, Dang P.V, Pennacchi P, Luca A.D and Flumian F 2017. Experimental evidence of a two-axial groove hydrodynamic journal bearing under severe operation conditions. *Tribology International* **109** 416-427.

[10] Dang P.V, Chatterton S, Pennacchi P 2019. The Effect of the Pivot Stiffness on the Performances of Five-Pad Tilting Pad Bearings. *Lubricants* **7**, 61.

Acknowledgement
This research is funded by Funds for Science and Technology Development of the University of Danang under project number B2019-DN02-67.

Multi Frequency on RGB Lighting LED Driver Technique

Jirawath Parnklang[1], Lalita Phikunthong[1], Rawisara Jantri[1], and Pichkanit Sukpasri[2]

[1] Department of Electronics Engineering, Faculty of Engineering, King Mongkut:s Institute of Technology Ladkrabang Bangkok, Thailand10520
[2] Energy Management Department, CP ALL Public Co., Ltd. CP Tower Silom Road, Bangkok, Thailand 10504

E-mail: jirawath.par@kmitl.ac.th; sonui270804@gmail.com

Abstract. In this paper, we conducted RGB incident light intensity of the high power white lighting LED. The current control 50% duty cycle pulse signal source with appropriate semiconductor carrier frequency response are supply to the RGB white lighting LED. Although, the carrier relaxation time of the red blue and green light in the white lighting LED are not the same value which depend on the carrier different of the energy level. The measurement show that the lowest duty cycle appropriate pulse signal for only blue or UV base white LED give more relative light intensity than normal (low frequency) signal within 11%.

1. Introduction

All type of white lighting LED has experienced substantial transformations in recent years. Light Emitting Diode (LED) technology is the first primary reason for this revolution. Fast white LED uptake has subsequently highlighted new possibilities for lighting system. Advances in COB LED chip technology allow for a shift in lighting function from purely brightening, to a human-centric approach. For the first time in the past, lighting design does not only just address the visual effect of illumination, but also can increasingly add the emotional and biological wellbeing of a person(worm white, cool white or daylight etc.), and even plants. Furthermore, the wider smart green smart city approaches enabling the integration of lighting into the wider urban landscape.
Nowadays, to mixed-color of white light: One technique is to mix the light from several colored LEDs to create a spectral power distribution that appears white light. Similarly, so-called tri-phosphor fluorescent lamps use three phosphors, each emitting a relatively narrow spectrum of blue, green or red light. By locating red, green and blue COB LEDs adjacent to one another, and properly mixing the amount of their output color, the resulting light is white in appearance. Phosphor-converted white light: Another approach to generating white light is by using of phosphors together with a short-wavelength LED. For example, when one phosphor material used in LEDs is illuminated by blue or ultra violet light, it emits yellow light having a fairly broad spectral power distribution. By incorporating the phosphor in the body of a blue LED with a peak wavelength around 450 to 470 nanometers, some of the blue light will be converted to yellow light (low power than blue light) by the phosphor. The remaining blue light, when mixed with the yellow light, results in white light. New phosphors are being developed to improve color rendering.

2. Theory

The basic P-N junction is the functioning of the today LED. The LED has an anode and a cathode separated by a crystal of semiconductor material. When a voltage is applied across the electrodes, the current flows from the anode (P side) to the cathode (N side). When an electron meets a hole at the P-N junction it falls in to a lower energy state to recombination with hole or falls to some energy level in the bandgap. The difference in energy of the two states is called the 'Band gap, which is a specific characteristic of the material comprising the P-N junction of the LED. There are two approaches to white light generation with LEDs: 1) phosphor based white LEDs, and 2) multi-colored LED solution – including RGB-LEDs. The RGB-LED solution has the advantages of variable color point and theoretically has higher efficiency than phosphor LEDs. As we know the red blue and green light LED must have the different energy band gap and form the data specification of white LED only show us about the spectrum of the composite RBG light to be a worm white, cool white or daylight white LED. The different constituents of visible light red has the longest wavelength of 700 nm (least energetic) while violet has the shortest wavelength 400 nm (most energetic)Since an LED produces light in a narrow band of wavelengths phosphors are often used to improve the spectrum of white light produced by an LED. It is also possible to combine several LEDs each producing a different wavelength to produce full spectrum light. Today white lighting LED is based on blue LED or UV LED. So it is easy to review the carrier relaxation times that laminate the white light.

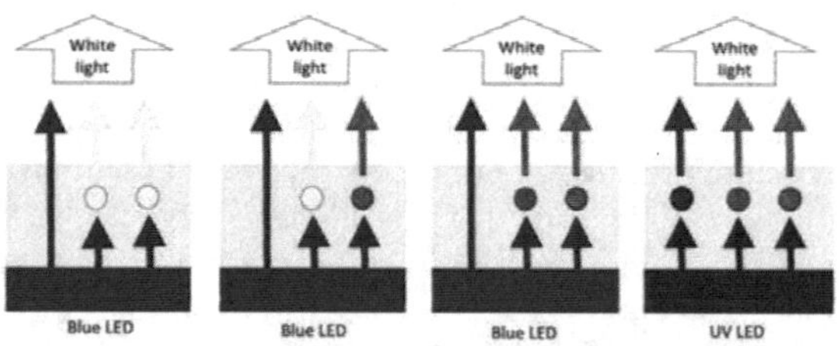

Figure 1. RGB white LED configuration [1].

For a white LED based lighting system, both red green blue (RGB) LEDs and phosphor based LEDs have been fabricated. Now, many of commercially available super bright white lighting LEDs are phosphor based, which consisting of blue or UV LEDs interference or filter or energy reduction with a yellow phosphor coating, which offers low cost and low complex. However, phosphor based LED show a limitation in the modulation bandwidth available due to the long relaxation time of the phosphor. The super bright lighting white LED's in commercial production today are based on an InGaN-GaN structure, and emit blue light. (wavelength range between 450 nm – 470 nm) The GaN based and InGaN active layer LEDs are changed by a yellow phosphor coating, that made of Cerium doped Yttrium Aluminum Garnet (Ce3+:YAG) crystals, which have been powdered and bound in a type of viscous adhesive.

A new type of blue LED was first demonstrated by the Cree, Inc. USA in September 2003 to give 24 mW. at 20 mA. CREE produced a commercial package of white lighting giving 65lm/w performance at 20 mA. This has been the brightest white LED with lumen power four times greater than commercially available white LED and more than four times more efficient than standard intensity. In 2006 CREE demonstrated a prototype white LED with 131 lm/W efficiency at 20 mA. Also Seoul Semiconductor, USA has plan for 135 lm/W by 2007 and 145 lm/W by 2008, which would be approaching reasonable order of magnitude improvement over standard and better even than any standard fluorescents. Nichia corp. Japan has developed a white light LED with efficiency of 150 lm/W at a forward current of 20 mA. In the next future it will be replacing various types of lighting accessories like incandescent bulbs, especially in every vehicle light. [2]

There are various carrier relaxation processes in semiconductors, and probably the most important one is the relaxation of carrier drift momentum, mv_d, which is the product of the effective mass m and the drift velocity v_d. In the absence of an external electric field, the free electrons in a semiconductor substance will move in random directions, so that the resultant velocity of free

electrons in any direction is equal to zero. By applying an external electric field on the specimen, the free electrons of the semiconductor acquire a constant resultant velocity in the opposite direction to that of the applied electric field. If the field is turned off suddenly at $t = 0$, then the average drift velocity $\langle v_x \rangle 0$ reduces exponentially to zero. The electron recombination time τ is called relaxation time. Due to the white led in figure 1, the newly carrier relaxation time are normally limit. Table 1 show some carrier relaxation time in semiconductor. [3][4][5]
We have been well know that the charge carriers in metals are electrons which are in partially occupied conduction band (*Ec*). Electrons in fully occupied valence bands (*Ev*) do not contribute to net current. We understand this bya consideration of the effective mass of electrons (m^*) near the top of the valence band. It may be seen that near the top of a band, the effective mass is negative (simple case). This means that in an electric field, the electrons which are near the top of the valence band are accelerated in the same direction of the field while those near the bottom are accelerated in the opposite direction. (this behavior is hole in semiconductor).In semiconductors, the width of the forbidden band is small enough so that electrons near the top of avalanche band can easily jump to the unoccupied conduction band. The semiconductor carrier relaxation time (τ) can be simple calculated from mobility (μ), electron charge (q) and the carrier mass (m) by equation (1)

$$\tau = \frac{m\mu}{q} \qquad (1)$$

Table1. Parameter values for the energy relaxation time Model for basic material.

Material	$\tau_{\epsilon,0}$ [ps]	$\tau_{\epsilon,1}$ [ps]	C_0	C_1	C_2	C_3	$\tau_{\epsilon,2}$ [ps]
Si	1.0	-0.538	0	0.0015	-0.09	0.17	0.4
Ge	0.26	1.49	0	-0.434	1.322	0	0.4
GaAs	0.48	0.025	0	-0.053	0.853	0.5	1.0
AlAs	0.17	0.025	61	-0.053	0.853	0.5	1.0
InAs	0.08	0.025	3	-0.053	0.853	0.5	1.0
InP	0.5	0.21	-18	-0.04	0	0	1.0
GaP	0.04	0.21	10	-0.04	0	0	1.0

3. Experimental

The first test are the light intensity relate to the frequency of each sample. The sample blue base white LED that use in this measurements are CREE Xlamp XM-L LEDs and 5050 White LED PKG. All of the sample are common cathode. In the final test, the power efficiency, we have already cut the electrode apart. Figure 2 show the intensity flux relate to the current of the sample. The forward voltage of this LED is 2.9V at 700 mA forward current. [5] [6] The RGB base white led sample is Everylight brand 5 mm Round White LED (T-1 3/4) (Forward Voltage V_F = 3.0 to 3.6 V I_F= 20mA [6][7].

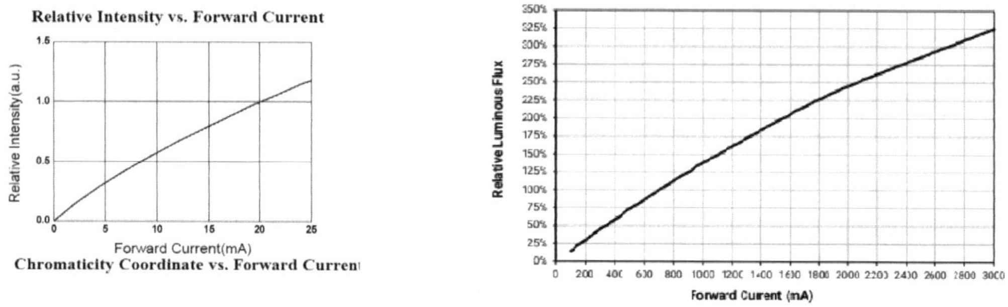

Figure 2. Relative Flux VS Current of CREE Xlamp XM-L LEDs and Everylight brand 5 mm Round white LED.

From the upper information, a 5.8Vpeak (V_{avg} = 2.9V) 700 mA. current controlled for CREE and 5050LED, 20 mA 7.2Vpeak for Everylight brand, 50% duty cycle time pulse with 2μS on/off time and also 1.5 μS, 2.5μS, 3μS, 3.5μS and 4μS are apply to the sample LEDs. The light intensity are measurement at 0 degree direction from the sample LEDs. The measurement system and experimental results show in figure 3. After that the 30%, 50% and 65% duty cycle time are applied to the blue, green and red LED with appropriate frequency. The measurement results are shown in figure 4 respectively.

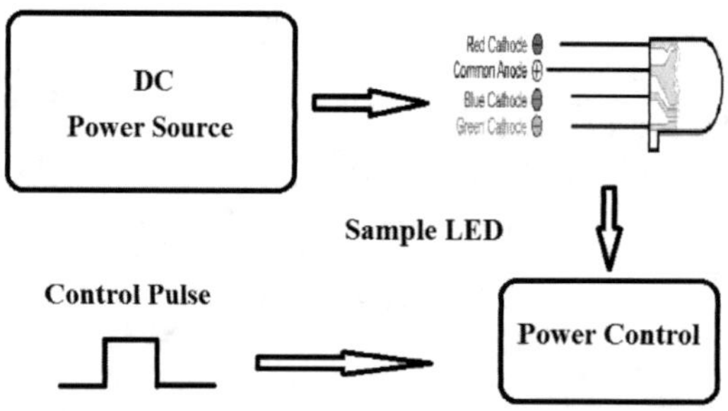

Figure 3. Light Intensity Testing Circuit System.

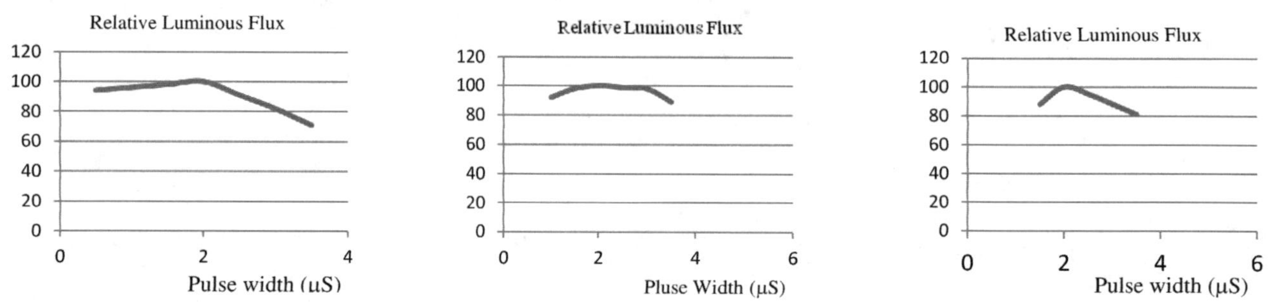

(a) (b) (c)

Figure 4. The experimental results of (a) CREE, (b) 5050LED and (c) Everylight LED and near UV base white LED.

After creating the LED light test system successfully, the LED sample are measured of light intensity at PTEC (Electrical and Electronic Product Testing Center) with standard tools and lighting equipment testing (Integrating Sphere).

Regardless of the results of the experiment, to fabricate a real driving signal for commercial work maybe the most difficult. Due to the high frequency of the pulse generator make it difficult to the PCB design and cannot be done at a cheap price. But experimenting with this concept may cause future development. Form the data sheet and the experimental results of figure 4, the 700mA supply current to CREE LEDs and the 5050LED create 260 lumen of white light. But raw data of the experimental results, the maximum luminous is 65 lumen. Due to the now high power white for lighting LED (not for communication LED) response frequency is very low. This sample LED maximum laboratory testing frequency is 1kH pulse. But the experiment data show that if the frequency response of the lighting LED is equal to the low power communication LED the up-close relaxation time pulse is the next driving signal. The maximum intensity of this device is the 2μS pulse width. Figure 4 show the similar response but the wide range of the data may mean that the white light are stimulate form multi energy level of the carrier in the indirect bandgap for the red blue and green energy level. The intensity to the power input efficiency of led is tested by control the input frequency of each LED color light. The maximum frequency is set to the blue and the lowest frequency is for the red. The measurement of the efficiency of the single low frequency is 11% lower than the multi frequency.
The light intensity from the experiment is very low maybe cause by the amount of the power obtained from the power source is low at limited time. The light carrier in semiconductor is too low. This should affect the light intensity. However, if the frequency response of the lighting LED is developed to increase, the experimental light intensity should be higher. In the future, the generate of high frequency pulse signal source is cheap and also the frequency response improvement of the high power white LED for lighting is suitable with high frequency pulse the lm/w of the LED light is come true.[8]

4. Conclusion
We have conducted sample LEDs experiment with the electrical measurement tools and very much limited. But the results of the experiments based on the ideas obtained are sufficient to summarize the data.
White light LED from blue light LED have a narrow band relaxation time. The peak luminous is at the 250 kHz. And white light LED from RGB base light LED have wider band relaxation time the blue base LED because the RGB light stimulate from various energy gap. This experiment, the appropriate pulse 5.8Vp 700 mA and 7.2 Vp 20 mA current control used in the lighting LED driver with 50% duty cycle signal. The lighting intensity LED has a maximum value luminous flux compared to LED power. The results showed that if the average frequency too much, luminous efficiency is decrease because high LED power and low frequency to make luminous flux resulting from peak relaxation time impact the low (but very high) total luminous flux. Therefore, the appropriate frequency should be made a suitable ratio between single τ and steady state constant current. The results found that the most appropriate pulse is not over 20 time more than relaxation time. When considering the simple relaxation time equation, one of the improvement is to control the m of the semiconductor carrier which stimulate to the excite energy level and reduce the energy by emit the blue light. This m parameter is different when the position of m in the energy band is different. Now the control position of m in the energy band cannot be done. Currently, finding the m position in excite band is just a statistical principle. The chance to find m in the energy band at the specified location with a probability of 50 percent. This is still the reason of white LED that the wide band spectrum of the light (Red Blue or Green) from the LED has a problem. In the future, if the spectrum width of the LED light is reduce and also the single crystal ingot of semiconductor is 100% perfect, the control of m value for better control of τ results.
Overall, the actual measurement frequency obtained from the experiment and get the maximum light intensity value is very similar. White LEDs have low UV content. The blue light radiation content is not much higher than the blue light radiation of the other white light sources. Due to the fact that almost commercial white LEDs work with such blue LEDs that excite their phosphors efficiently (i.e. the blue peak wavelength is in the range of 450 nm to 460 nm), they avoid some characteristic fluorescent short wavelength maximum (404.7 nm, 407.8 nm and also 435.8 nm). And that this experiment is still interested in creating white light from red, green, blue light because currently there is

a group of commercial LED lamps of this type. And experimental results still in accordance with the concept of the theory in all respects. In general LED driver circuit designers are often unaware that the LED devices that are developed is a direct or indirect bandgap. The gap is direct or indirect is important in the application of inorganic semiconductors to photoconduction and electroluminescence, because it is one of the major factors that determines the efficiency of these processes. But the circuit designer only choose the most efficient LED as the manufacturer provides. This is a general problem of driver circuit design. The author hopes the results of this experiment will be an important way to develop future LED driver circuit design

5. Reference

[1] Takashi Takeda, Rong- Jun Xie, Takayuki Suehiro, Naoto Hirosaki. Nitride and oxynitride phosphors for white LEDs: Synthesis, new phosphor discovery, crystal structure. *Progress in Solid State ChemistryVolume 51, September 2018, Pages 41-51.* DOI= http://doi.acm.org/10.1016/j.progsolidstchem.2017.11.002.

[2] J. K. Singh1, B.R .Singh 2, Kuldip Singh2 and V Kumar 3, Fabrication of High Brightness Blue LED, International Journal of Applied Engineering Research, ISSN 0973-4562 Volume 5 Number 2 (2010) pp. 247–252© Research India Publicationshttp://www.ripublication.com/ijaer.htm.

[3] Vassil Palankovski, Dissertation Simulation of Heterojunction of Bipolar transistor http://www.iue.tuwien.ac.at/phd/palankovski/node51.html.

[4] Osram Opto Semiconductor, "Electrical Charac-teristics of LEDs," Application Note 3256, OsramOpto emiconductors.

[5] CREE Product family data sheet .CLD DS33REV 11V. https://www.cree.com/led-components/media/documents/XLampXML-11E.pdf

[6] P. H. Binh, V. D. Trong, P. Renucci and X. Marie." A simple sub-nanosecond ultraviolet light pulse generator with high repetition rate and power, " Institute of Materials Science, Vietnam Academy of Science and Technology, Vietnam, August 2013

[7] Joachim Rose, Stella Bradbury, Isabel Bond, Paul Ogden, Andrew Price, Richard Oliver and Yerbo lKhassen. "Driving LED in a Nanosecond Regime by a Fast Operational Amplifier," Department of Physics and Astronomy, Faculty of Mathematics and Physical Sciences, University of Leeds, LS2 9JT, 31 July 2009.

[8] Hansen, E. (2015, November 24). Lighting designprofessor at Aalborg University. Personal communications(S. Getova, C. Pai & S. Soemadiredja, Interviewers) Ilkem Ozge Huyal, Tuncay Ozel, Sedat Nizamoglu, Unsal Koldemir, Donus Tuncel, and Hilmi Volkan Demir, White Light Generation with Azide Functionalized Polyfluorene

Acknowledgement

The authors are grateful to National Research Council of Thailand (NRCT), all of the Faille Analysis Team of the RF laboratory of Armstrong Industrial Thailand Co.LTD and also department of Electronic Engineering Faculty of Engineering, King Mongkut's Institute of Technology Ladkrabang for their official support and encouragements. Lastly, thank you to the engineer, Luminous Intensity Laboratory, PTEC, for supporting the experiment.

IOP Conf. Series: Materials Science and Engineering **895** (2020) 012011 doi:10.1088/1757-899X/895/1/012011

Research and development of a trajectory-position controller for an underwater glider

Boris Gurenko, Kiril Surov, Anatoly Nazarkin

Southern Federal University, Rostov-on-Don, Russia

boris.gurenko@gmail.com

Abstract. In this article, development of glider control system for movement in a three-dimensional based on position-trajectory control laws is considered. To ensure control, a model of spatial motion of the glider is used, taking into account the component of the variable buoyancy. As a result of the simulation, the trajectories of the glider were obtained at points set in the horizontal plane, considering the parameters of the trajectory for movement in the vertical plane. The results obtained make it possible to evaluate the adequacy of the developed control system. The deviation of the system when the glider reaches the set point is minimal, and using the trajectory parameters in the vertical plane, it is possible to control the depth of immersion and speed. The presented results of the simulation of coating elements prove the possibility of practical use of the control system when planning glider-type AUV missions.

1. Introduction

In this work we review the synthesis of a controller for a glider-type underwater vehicle based on a position-trajectory controller described in [1, 2]. This controller is widely used for autonomous control of various vehicles [3].

The main advantages of this controller make it possible to control non-linear objects that consider external environmental disturbances as well as consider interconnected and non-linear mathematical models of a plant.

2. Underwater glider technology

An underwater glider is an autonomous underwater vehicle that uses variable-buoyancy device for its propulsion. Compared to conventional submarine-type vehicles that employ propellers, gliders dive and float up alternately that, with use of underwater wings (hydrofoils), causes the vehicle to follow a sinewave or sawtooth-like trajectory. While gliders are much slower than most submarines, they offer significantly more range, making them suitable for various continuous research missions, lasting weeks or even months and covering thousands of kilometers of range.

The variable-buoyancy system usually consists of two parts: ballast reservoir and center of mass displacement device. The former allows to change the overall buoyancy of the vehicle, causing it to dive or float up, and the latter moves the center of vehicle's mass, thus creating pitch angle. The above principles force the glider to follow the signature sawtooth trajectory, propelling it forward.

However, in case of the vehicle model we're reviewing in this paper, the mass displacement device is excluded. Thereby, the ballast reservoir has to be located in a compartment located in the bow of the vehicle. In this case, the reservoir combines the two glider propulsion principles described above. In empty state, the glider is floating on water surface with slightly positive pitch angle (with its nose up).

IOP Conf. Series: Materials Science and Engineering **895** (2020) 012011 doi:10.1088/1757-899X/895/1/012011

As the reservoir is getting filled, the buoyancy reduces and the center of mass moves in direction of the glider's bow. That forces the vehicle to dive and move forward. As the reservoir is getting emptied, the process reverses and the glider floats up.
The proposed scheme may reduce the power consumption of onboard systems even more, greatly extending mission time and overall durability and endurance. In order to perform turns, we propose to split the ballast compartment into two parts, applied for left and right turns respectively.

3. Mathematical model definition

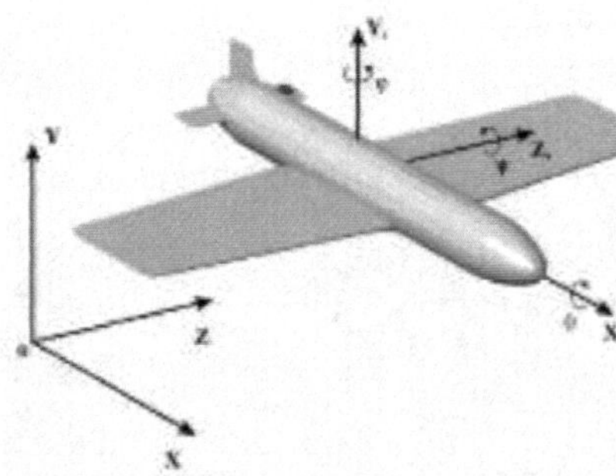

Figure 1. Coordinate systems and positive directions of Euler angles.

The cited works [4, 5] present the structure of a mathematical model that considers the glider's movement. To derive the mathematical model of a glider, we use the following rectangular coordinate systems, shown in Figure 1.
In general, the mathematical model of the glider's motion in a three-dimensional environment may be represented by Eq. (1.1–1.6):
Where p_w is water density; V is displacement; v_{x1}, v_{y1}, v_{z1} are projections of object's velocity on axis OX1, OY1, OZ1 (or surge, heave and sway respectively); v_0 is the absolute value of velocity; ω_{x1}, ω_{y1}, ω_{z1} are angular velocities around axes OX1, OY1, OZ1; J_{x1}, J_{y1}, J_{z1} are moments of inertia relative to axis OX1, OY1, OZ1; $\lambda_{11}, \lambda_{22}, \lambda_{33}$ are added masses of glider's hull, $\lambda_{26}, \lambda_{35}$ are added static moments of the hull; $\lambda_{44}, \lambda_{55}, \lambda_{66}$ are added moments of inertia; c_{x1}, c_{y1}, m_{z1} are coefficients of positional hydrodynamic forces and moments; $c_{y1}^{\omega z1}$, $m_{z1}^{\omega z1}$ are coefficients of rotational hydrodynamic forces and moments; p is excess buoyancy; x_p, y_p, z_p are the arms of excess buoyancy in glider's principal axes; $c_{z1}^{\beta}, m_{x1}^{\beta}, m_{y1}^{\beta}$ are coefficients of derivative positional hydrodynamic forces and moments; $c_{z1}^{\omega_{z1}}, m_{x1}^{\omega_{x1}}, m_{x1}^{\omega_{y1}}, m_{y1}^{\omega_{x1}}, m_{y1}^{\omega_{y1}}$ are coefficients of derivative rotational forces and moments; h is metacentric height; ψ is pitch angle; θ is roll angle; φ is yaw angle.

$$(p_w V + \lambda_{11})\dot{v}_{x1} = c_{x1}\frac{p_w V_0^2}{2}V^{2/3} + (p_w V + \lambda_{22})V_{y1}\omega_{x1} + \lambda_{26}\omega_{z1}^2 + +p\sin\psi; \quad (1.1)$$

$$(p_W V + \lambda_{22})\dot{v}_{y1} + \lambda_{26}\dot{\omega}_{z1} = c_{y1}\frac{p_w v_0^2}{2}V^{2/3} + c_{y1}^{\omega_{z1}}\frac{p_w v_0}{2}\omega_{z1}V - (p_w V + +\lambda_{11})v_{x1}\omega_{z1} + +p\cos\psi; \quad (1.2)$$

$$(J_{z1} + \lambda_{66})\dot{\omega}_{z1} + \lambda_{26}\dot{v}_{y1} = m_{z1}\frac{p_w v_0^2}{2}V + m_{z1}^{\omega_{z1}}\frac{p_w v_0}{2}\omega_{z1}V^{4/3} - -p_w V g h\sin\psi - \lambda_{26}v_{x1}\omega_{z1} + +p(x_p\cos\psi - y_p\sin\psi); \quad (1.3)$$

$$(p_w V + \lambda_{33})\dot{v}_{z1} + \lambda_{35}\dot{\omega}_{y1} = c_{z1}^{\beta}\frac{p_w v_0}{2}V^{2/3}v_{z1} + c_{z1}^{\omega_{z1}}\frac{p_w v_0}{2}\omega_{y1}V - -\omega_{x1}(p_w V + \lambda_{22})v_{y1} + +\omega_{y1}(p_w V + \lambda_{11})v_{x1} - p\cos\psi\sin\theta; \quad (1.4)$$

$$(J_{x1} + \lambda_{44})\dot{\omega}_{x1} + \omega_{y1}\lambda_{26}v_{y1} + v_{y1}\lambda_{35}\omega_{y1} = m_{x1}^{\beta}\frac{p_w v_0}{2}V v_{z1} + +m_{x1}^{\omega_{x1}}\frac{p_w v_0}{2}V^{4/3}\omega_{x1} + m_{x1}^{\omega_{y1}}\frac{p_w v_0}{2}V^{4/3}\omega_{y1} - p_w g V h\sin\theta\cos\psi - p(z_p\cos\theta + +y_p\sin\theta)\cos\psi; \quad (1.5)$$

$$(J_{y1} + \lambda_{55})\dot{\omega}_{y1} + \lambda_{35}\dot{v}_{z1} = m_{y1}^{\beta}\frac{p_w v_0}{2}V v_{z1} + m_{y1}^{\omega_{x1}}\frac{p_w v_0}{2}\omega_{x1}V^{4/3} + +m_{y1}^{\omega_{y1}}\frac{p_w v_0}{2}\omega_{y1}V^{4/3} + \omega_{x1}\lambda_{26}v_{y1} + v_{x1}\lambda_{35}\omega_{y1} + p(x_p\cos\psi\sin\theta + z_p\sin\psi). \quad (1.6)$$

Coefficients p and z_p are used as main control parameters. We synthesize the controller according to position-trajectory control law considering changing the arm length of excess buoyancy z_p. The above control parameters and their directions are shown in Figure 2.

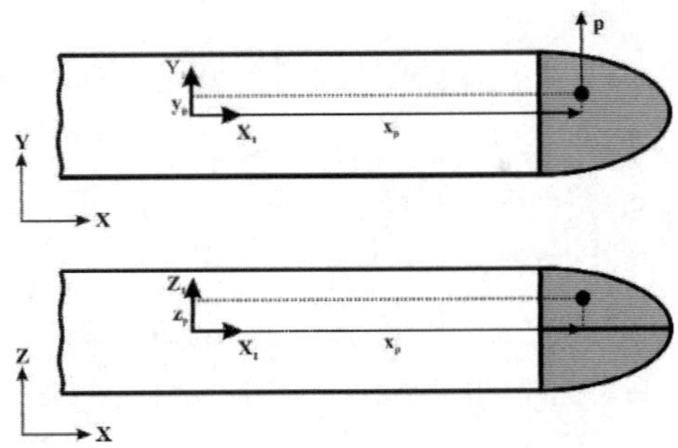

Figure 2. Coordinate systems and control parameters.

4. Control system synthesis

The process of developing a controller for an underwater glider has a few main steps:

Defining the target path function y(x) and z(x). That function must represent the target of control as an error of close loop system;

Defining the trajectory error based on difference between coordinates of real and desired trajectories;

Evaluating control parameters p and z_p.

To solve the problem of synthesizing the control system we will divide it into two parts:

Control system for moving along the plane OX_1Y_1 which matches surge and heave motions;

Control system for moving along the plane OX_1Z_1 which matches the sway motion and turning.

We define an error of closed-loop system along the plane OX_1Y_1 (2):

$$e_y = y - y_0, \tag{2}$$

Where y_0 is function of a desired trajectory, $y_0 = A\sin(\omega t) + A_m$, where A is dive amplitude, ω is dive frequency, A_m is average depth of trajectory.

By substituting the desired path function into (2), we get the value of an error (3):

$$e_y = y - A\sin(\omega t) - A_m. \tag{3}$$

According to the position-trajectory method of control, we introduce an equation that satisfies the condition of asymptotic stability (4):

$$\ddot{e_y} + a_1\dot{e_y} + a_2 e_y = 0, \tag{4}$$

Where e_y is an error that includes the difference between real and desired trajectories along the plane OX_1Y_1; a_1, a_2 are tuning parameters of the controller.

Due to control being implemented by change of buoyancy, we need to express the excess buoyancy function p from equation (4).

By deriving eq. (1.1 – 1.6) and substituting the required parameters from mathematical model into eq. (4), it is possible to express the value of variable buoyancy p (5):

$$P_{num} = -A\omega^2\sin(\omega t) - a_1\big[(a_{12}v_{x1} + a_{22}v_{y1} + a_{32}v_{z1}) - A\omega\cos(\omega t)\big] - a_2[y -$$
$$-A\sin(\omega t) + A_m] - \frac{a_{12}A1}{\rho_w V + \lambda_{11}} - a_{22}\left[\frac{J_{z1}+\lambda_{66}}{\det O}A2 + \frac{-\lambda_{26}}{\det O}A3\right] - a_{32}\left[\frac{J_{y1}+\lambda_{22}}{\det O_2}(A4 - A5) +\right.$$
$$\left. + \frac{-\lambda_{35}}{\det O_2}(A6 + A7)\right] - \dot{a}_{12}v_{x1} - \dot{a}_{22}v_{y1} - \dot{a}_{32}v_{z1};$$

$$P_{denom} =$$
$$\frac{\cos\psi}{\rho_w V + \lambda_{11}} + a_{22}\left[\frac{J_{z1}+\lambda_{66}}{\det O}\cos\psi + \frac{-\lambda_{26}}{\det O}(x_p\cos\psi - y_p\sin\psi)\right] - -a_{32}\left[\frac{J_{y1}+\lambda_{55}}{\det O_2}\cos\psi\sin\theta -\right.$$
$$\left.\frac{-\lambda_{35}}{\det O_2}(x_p\cos\psi\sin\theta - z_p\sin\psi)\right];$$

$$p = {}^{P_{num}}/_{P_{denom}}, \quad (5)$$

where $a_{12} = \sin\psi$, $a_{22} = \cos\psi\cos\theta$, $a_{32} = -\cos\psi\sin\theta$, $\dot{a}_{12} = \cos\psi\,(\omega_{z1}\cos\theta + \omega_{y1}\sin\theta)$,
$\dot{a}_{22} = -\sin\psi\cos\theta\,(\omega_{z1}\cos\theta + \omega_{y1}\sin\theta) - \cos\psi\sin\theta\,(\omega_{x1} - -\tan\psi\,(\omega_{x1}\cos\theta - \omega_{z1}\sin\theta))$;
$\dot{a}_{22} = \sin\psi\sin\theta\,(\omega_{z1}\cos\theta + \omega_{y1}\sin\theta) - \cos\psi\cos\theta\,(\omega_{x1} - -\tan\psi\,(\omega_{x1}\cos\theta - \omega_{z1}\sin\theta))$; $\det O = (\rho_w V + \lambda_{22})(J_{z1} + \lambda_{66}) - {\lambda_{26}}^2$;
$\det O_2 = (\rho_w V + \lambda_{11})(J_{y1} + \lambda_{55}) - {\lambda_{35}}^2$;
$A1 = c_{x1}\frac{p_w V_0^2}{2}V^{2/3} + (p_w V + \lambda_{22})V_{y1}\omega_{x1} + \lambda_{26}\omega_{z1}^2$;
$A2 = c_{y1}\frac{p_w v_0^2}{2}V^{2/3} + c_{y1}^{\omega_{z1}}\frac{p_w v_0}{2}\omega_{z1}V - (p_w V + \lambda_{11})v_{x1}\omega_{z1}$;
$A3 = m_{z1}\frac{p_w v_0^2}{2}V + m_{z1}^{\omega_{z1}}\frac{p_w v_0}{2}\omega_{z1}V^{4/3} - p_w V g h \sin\psi - \lambda_{26}v_{x1}\omega_{z1}$;
$A4 = c_{z1}^{\beta}\frac{p_w v_0}{2}V^{2/3}v_{z1} + c_{z1}^{\omega_{z1}}\frac{p_w v_0}{2}\omega_{y1}V$; $A5 = \omega_{x1}(p_w V + \lambda_{22})v_{y1} + \omega_{y1}(p_w V + \lambda_{11})v_{x1}$;
$A6 = m_{y1}^{\beta}\frac{p_w v_0}{2}V v_{z1} + m_{y1}^{\omega_{x1}}\frac{p_w v_0}{2}\omega_{x1}V^{4/3} + m_{y1}^{\omega_{y1}}\frac{p_w v_0}{2}\omega_{y1}V^{4/3}$;
$A7 = \omega_{x1}\lambda_{26}v_{y1} + v_{x1}\lambda_{35}\omega_{y1}$
are shortened expressions introduced to simplify the calculation.

Next, we will define the control system for maneuvering. In that case, the goal of control system is to provide movement through a set of defined points. In order to create the roll motion, we will change the z_p parameter that allows glider to turn. That principle can be used for control system of glider in coordinate plane OX_1Z_1. We define the desired path as a straight line, described as a linear equation (6):

$$z = kx + b, \quad (6)$$

Where k is a slope, b is an intercept (for simplicity we will define it as 0, in that case the line will run through the origin).

As known, the slope is $\frac{\Delta z}{\Delta x}$. Assume k as desired slope and k_s as real:

$$k_s = \frac{\Delta z}{\Delta x} = \frac{\dot{z}}{\dot{x}} = \frac{(a_{13}v_{x1} + a_{23}v_{y1} + a_{33}v_{z1})}{(a_{11}v_{x1} + a_{21}v_{y1} + a_{31}v_{z1})}, \quad (7)$$

Where $a_{11} = \cos\varphi\cos\psi$, $a_{13} = -\sin\varphi\cos\psi$; $a_{21} = \sin\varphi\sin\theta - \cos\varphi\sin\psi\cos\theta$, $a_{23} = \cos\varphi\sin\theta + \cos\varphi\sin\psi\sin\theta$, $a_{31} = \sin\varphi\cos\theta + \cos\varphi\sin\psi\sin\theta$, $a_{33} = \cos\varphi\sin\theta - \sin\varphi\sin\psi\sin\theta$.

Next, we will define the desired slope k. Because the input of the control system will accept the coordinates of a point, we define the slope k as (8):

$$k = \frac{\Delta z}{\Delta x} = \frac{z_1 - z_0}{x_1 - x_0}, \quad (8)$$

Where x_1, z_1 are the coordinates of a set point, x_0, z_0 are the current glider's coordinates.

These values were obtained by integrating equations (1). With use of this method, the slope k will be constantly recalculated, which will allow the control system to steer the glider more precisely.

Next, we'll define the error of control system in coordinate plane OX_1Z_1 (9):

$$e_z = k_s - k. \quad (9)$$

Substituting eq. (7) and (8), we get the error equation (10):

$$e_z = \frac{(a_{13}v_{x1} + a_{23}v_{y1} + a_{33}v_{z1})}{(a_{11}v_{x1} + a_{21}v_{y1} + a_{31}v_{z1})} - \frac{z_1 - z_0}{x_1 - x_0}. \quad (10)$$

From modelling the glider's motion in works [5], it is known that the best stability of a glider with attributes used in this research is achieved by a change of z_p parameter in the limits $-0{,}445 < z_p < +0{,}445$, where the negative values provide the left roll and positive values are the right roll. Therefore, if the glider is passing to the left from the desired path, the values of the error are negative and otherwise if passing to the right. Accordingly, the values of z_p parameter can be defined as system (11):

$$\begin{cases} z_p = 0 \; if \; e_z = 0; \\ z_p = 0{,}445 \; if \; e_z > 0; \\ z_p = -0{,}445 \; if \; e_z < 0. \end{cases} \qquad (11)$$

To make the tuning of z_p smoother, we use a logistic function (sigmoid curve) $f(x) = \frac{L}{1+e^{-k\,(x-x_0)}}$, where L is the curve's maximum value, x_0 the x-value of the curve's midpoint,
K is the steepness of the curve. In case of the control system, function will be defined as (12):

$$z_p = 0{,}445\left(\frac{2}{1+e^{-b\,e_z}} - 1\right), \qquad (12)$$

Where b is the steepness of the curve. The z_p-e_z graph with $b = 20$ is shown in Figure 3.

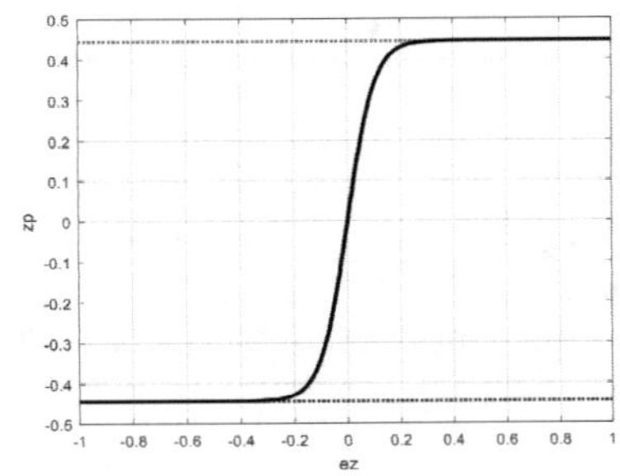

Figure 3. The z_p-e_z graph.

5. Research of the control system

For researching the synthesized controller, we use attributes of the Neptune glider [1, 6, 7]. According to the variable buoyancy control law (5), we'll consider an example simulation of longitudinal motion along the coordinate plane OX_1 Y_1. Using the function (5) in motion model 1, we will obtain the following motion graphs, shown in Figure 4.

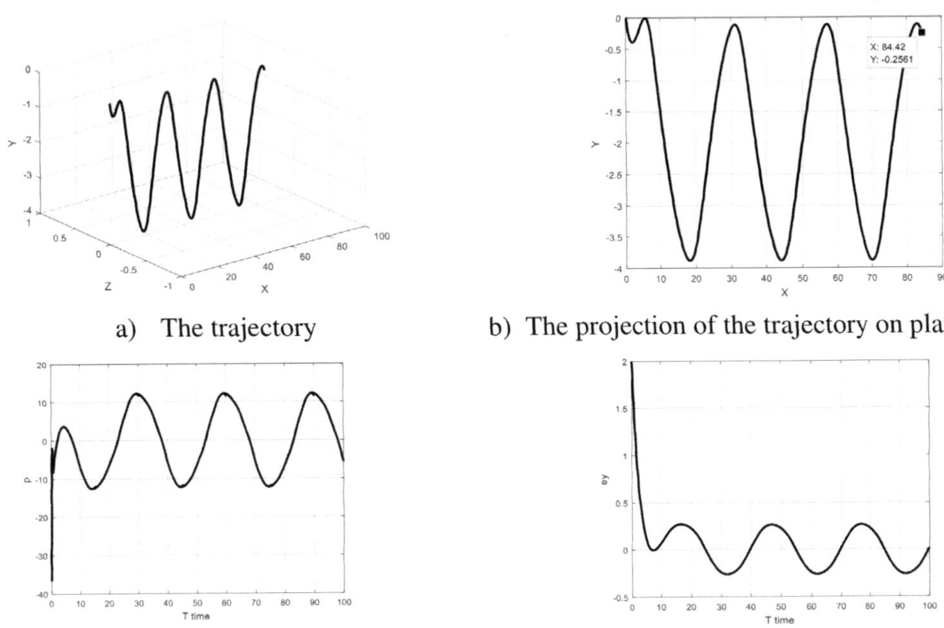

a) The trajectory

b) The projection of the trajectory on plane OXY

c) The value of buoyancy p

d) The value of error e_y

Figure 4. Graphs of various glider parameters.

IOP Conf. Series: Materials Science and Engineering **895** (2020) 012011 doi:10.1088/1757-899X/895/1/012011

For simulating the trajectory above we did not consider the tuning of z_p, its value was set to 0. We have chosen the following parameters of the sine wave: amplitude $A = 2$ meters, average depth A_m =2 meters, period $T = 30$ seconds. In 100 seconds, glider makes 84 meters with average speed $v_{x1} = 0{,}84\ m/s$ and does 3 full cycles of diving/ascending. Error e_y, after establishing the set mode, does not exceed 0.26 meters modulo.
The control system allows to change the depth, amplitude and frequency of dives, as shown in Figure5.

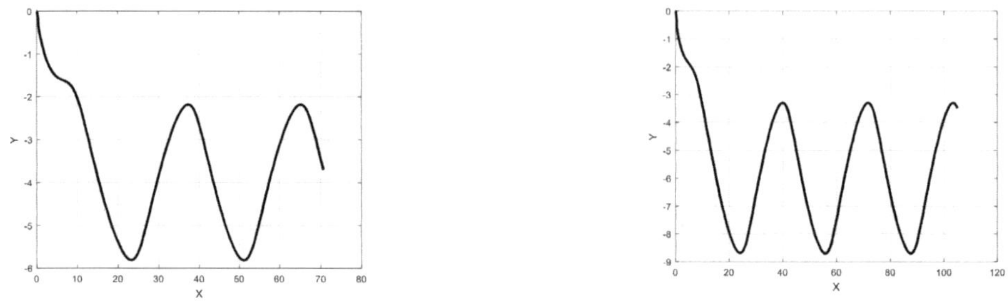

a) The trajectory with average depth $A_m = 4$ m, amplitude $A = 2$ m and period $T = 40$ s

b) The trajectory with average depth $A_m = 6$ m, amplitude $A = 2$ m and period $T = 30$ s

Figure 5. Glider trajectories with different parameters.

As seen in Figure 4, in the second case, glider travels a greater distance in the same amount of time. Consider the movement taking into account the control over the course using the z_p parameter. We will set the end point as (100; -70). Sine wave parameters are $A_m = 4\,m$, $A = 2\,m$, $T = 40\,s$. Simulation results are shown in Figure 6.

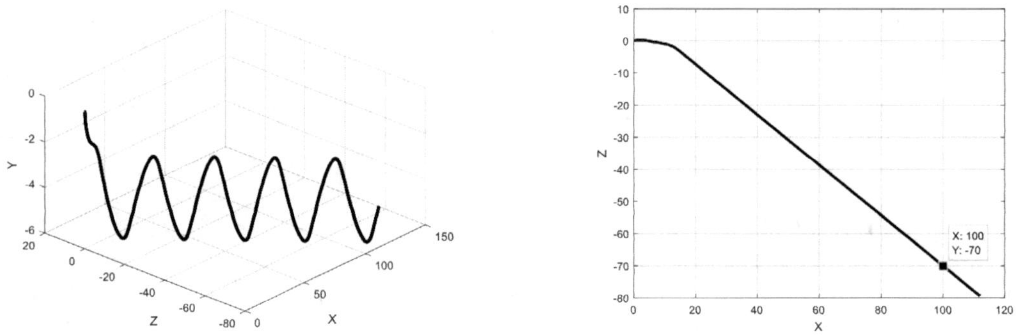

a) the trajectory

b) the projection of trajectory on plane OXZ

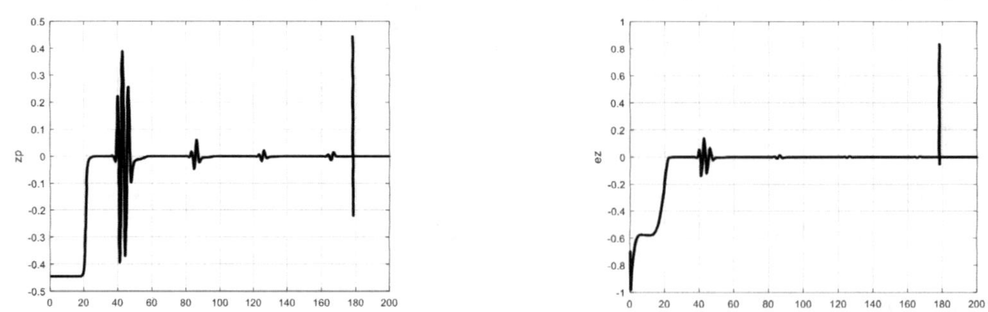

c) the value of z_p over time

d) the value of error e_z

Figure 6. Graphs of trajectory through point (100; -70).

IOP Conf. Series: Materials Science and Engineering **895** (2020) 012011 doi:10.1088/1757-899X/895/1/012011

The proposed control system allows to set several points for glider to move through. In the next example 3 points were set: (50; -20), (150; -30), (200; 0). Graphs of the trajectory is shown in Figure 7.

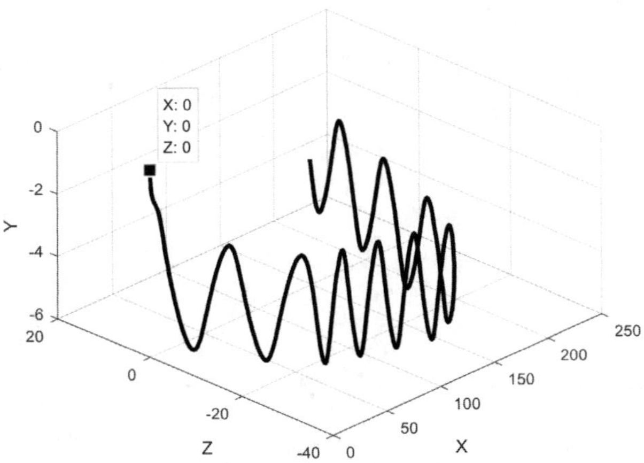

a) the trajectory

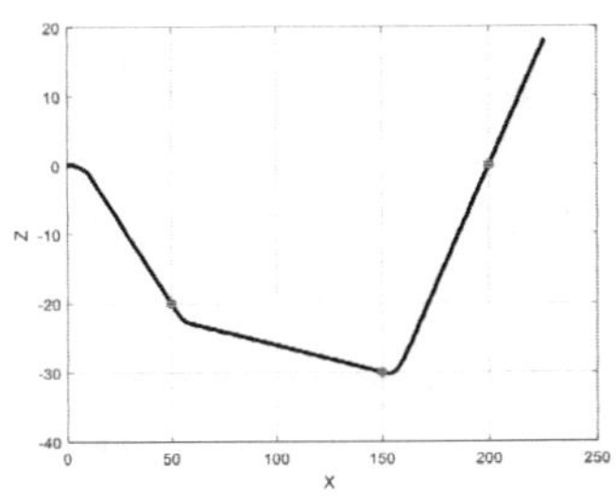

b) The projection of trajectory on plane OXZ (set points are marked)

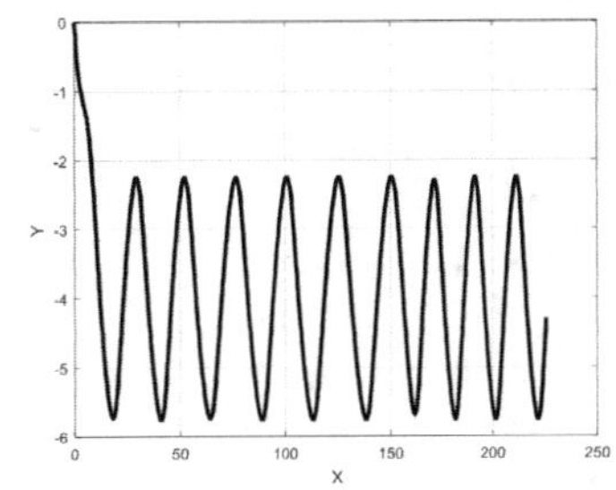

c) The projection of trajectory on plane OXY

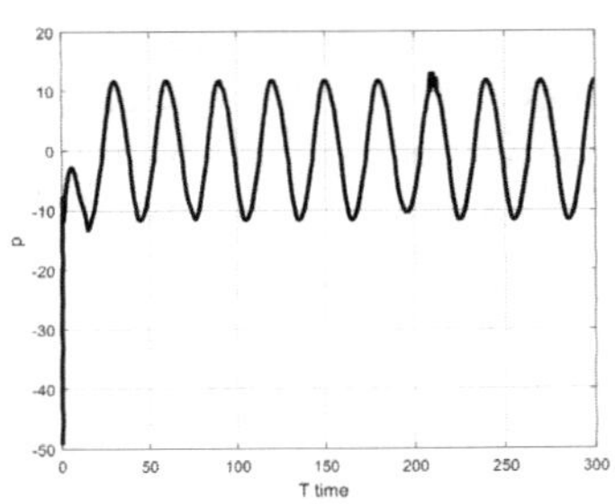

d) The value of buoyancy p

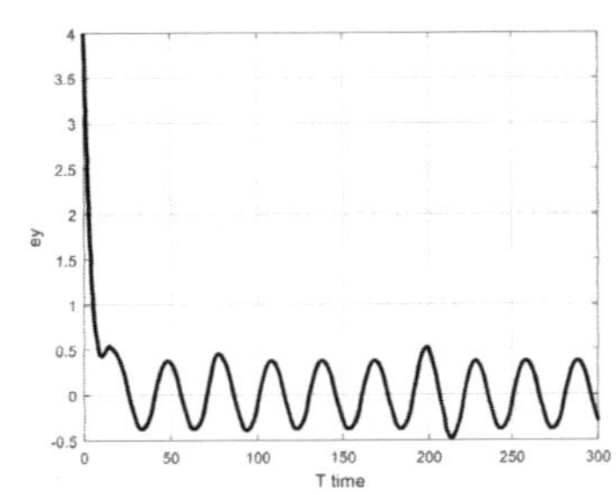

e) The value of error e_y

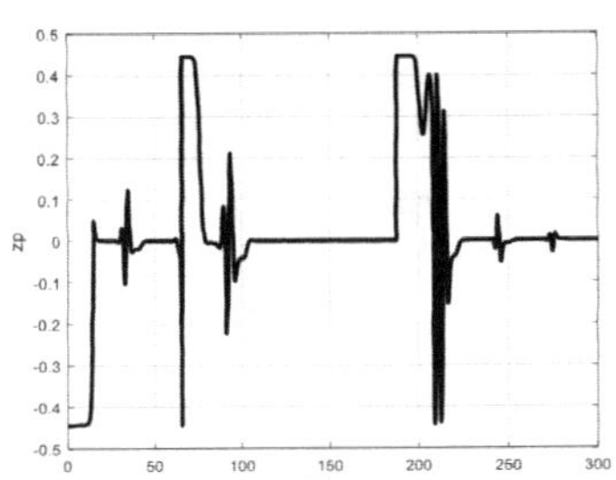

f) The value of z_p

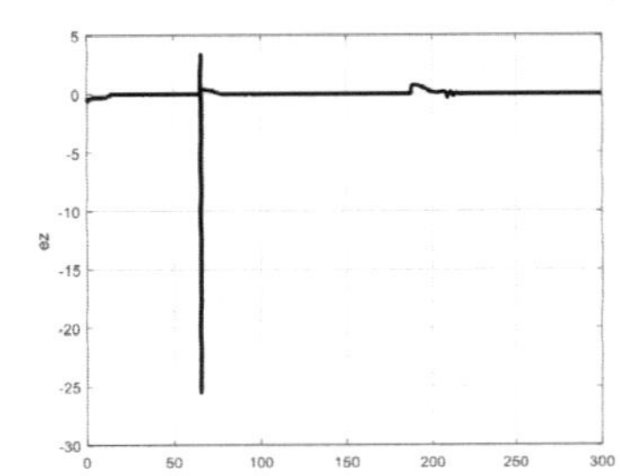

g) The value of error e_z

Figure 7. Graphs of trajectory via 3 points.

The most prevalent types of tasks for AUVs include search and inspections. Such objectives are conducted by surveys of selected areas of bottom surface and subsequent research of discovered objects. Usually, a typical mission of an AUV consists of area and point surveys. In order to conduct such surveys, the trajectory of a glider can be set as one of the special shapes, for example, meander or zigzag [8].
The meander shape provides an optimal coverage of researched area, because it does not contain overlaps and returns. An example of a meander is shown in Figure 8.

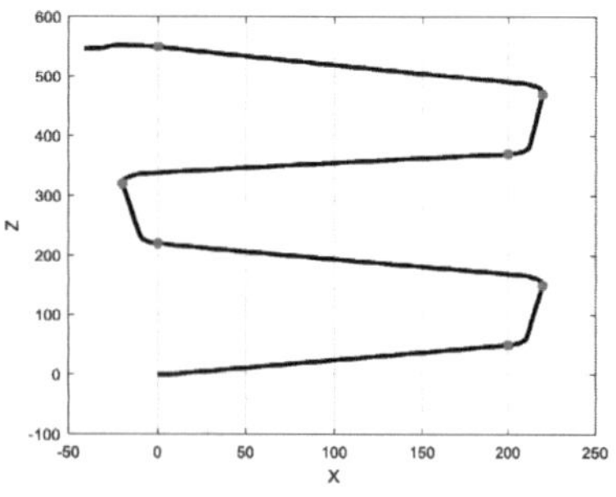

Figure 8. A meander-like shape.

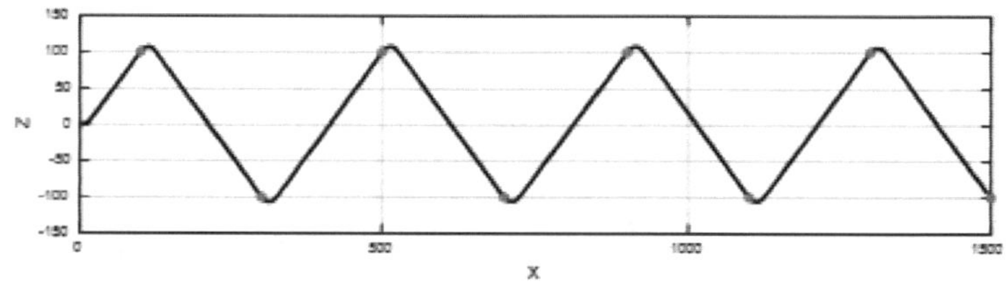

Figure 9. A zigzag shape.

If it is necessary to examine underwater infrastructure, such as cables and tubes, a zigzag trajectory can be used. An example is shown in Figure 9.
The developed control system can follow the shapes above, which makes it possible to conduct various underwater curves.

6. Conclusions
To sum up our work, the trajectories of the glider were obtained at points set in the horizontal plane, taking into account the parameters of the trajectory for movement in the vertical plane. The results obtained make it possible to evaluate the adequacy of the developed control system. The deviation of the system when the glider reaches the set point is minimal, and using the trajectory parameters in the vertical plane, it is possible to control the depth of immersion and speed. The above results of the simulation of coating elements prove the possibility of practical use of the control system when planning glider-type AUV missions. The simulation results are similar to the works [9, 10] and confirm the possibility of using the circuit without the center of mass displacement apparatus.

7. References
[1] Maevskiy A.M, Kostyukov V.A. Shkola-seminar «Molodezhno-studencheskie razrabotki i tvorcheskie proekty dlya issledovaniya i osvoeniya pribrezhno-shel'fovykh zon yuga Rossii». (Proc. 7th school-seminar "Student research projects and creative projects for researching and developing coastal areas of south of Russia") Gelendzhik, 2016, pp. 150-155.
[2] Pshikhopov V.Kh. Pozitsionno-traektornoe upravlenie podvizhnymi ob"ektami. [Position-trajectory control of moveable objects] Taganrog: TTI YuFU, 2009. 183 pp.
[3] Pshikhopov V.Kh., Fedotov A.A., Medvedev M.Yu., Medvedeva T.N., Gurenko B.V. Cyberleninka.ru URL: cyberleninka.ru/article/n/pozitsionno-traektornaya-sistema-pryamogo-adaptivnogo-upravleniya-morskimi-podvizhnymi-obektami
[4] Sukhorukov A.L., Titov M.A. Fundamental'naya i prikladnaya gidrofizika. 2012. №4. pp. 88-101.
[5] Sukhorukov A.L., Titov M.A. Fundamental'naya i prikladnaya gidrofizika. 2016. №3. pp. 66-79.
[6] Gurenko B.V., Nazarkin A.S. Inženernyj vestnik Dona (Rus). 2015. №4. URL: ivdon.ru/ru/magazine/archive/n4y2015/3288
[7] Gurenko B.V. Inženernyj vestnik Dona (Rus). 2014. №4. URL: ivdon.ru/ru/magazine/archive/N4y2014/2626

[8] Bagnitskiy A.V., Inzartsev A.V. Podvodnye issledovaniya i robototekhnika. 2010. №2. pp. 17-24.
[9] J. G. Graver. Underwater gliders: dynamics, control and design: PhD thesis in mechanical and aerospace engineering. Princeton University. 2005. 273 pp.
[10] N. Mahmoudian, C. Woolsey Underwater glider motion control. 47th IEEE Conference on Decision and Control. 2008. pp. 552-557.

Acknowledgments
The reported study was funded by RFBR according to the research project № 18-38-00441 (regarding the development of a position-trajectory regulator) and with the support of the grant of the President of the Russian Federation for young scientists MK-3099.2019.8 (regarding the study of movement in the Matlab environment).

IOP Conf. Series: Materials Science and Engineering **895** (2020) 012012 doi:10.1088/1757-899X/895/1/012012

Detection of Defects on Single-Bead Welding by Machine Learning Methods

Mucahid Barstugan[1,*], Yavuz Selim Ceran[2], Musa Yilmaz[2], Niyazi Adem Dundar[2]

[1]Electric and Electronics Engineering Department, Engineering and Natural Sciences Faculty, Konya Technical University
[2]R&D Center, Hürsan Hydraulic Press and Machine Tools Industry and Trade Inc. Konya, Turkey

*mbarstugan@ktun.edu.tr

Abstract. This study classified the defects that were encountered on single-bead welding, which was made by MIG/MAG welding machines in Hürsan Press company. 2250 images were taken on weldings that were made by five different welders. This study classified defects into three classes as good welding, porosity, and discontinuities. The images in the dataset, which have three classes, were classified by two stages. In the first stage, the texture features of the welding area were extracted. In the second stage, Artificial Neural Networks (ANN) method classified the extracted features. Sensitivity, specificity, accuracy, precision, and F-score metrics were used to measure the classification performance. 1500 images were used to train the system and training and validation performances were obtained as 94.03% and 94.19%, respectively. 750 images were used to test the performance of the proposed method and the test performance was obtained as 94.31%. The proposed method detected a defect on single-bead welding in 0.98 seconds.

1. Introduction

The manual process of welding has more defects than automatic welding. The defect occurrence probability changes according to the ability of the welder. The welding defects are detected by Non-destructive Testing (NDT) experts under enough light. Therefore, there is a probability of a wrong decision on the type of defect during the check process. The aim of this study is to assist NDT experts during welding check process. Also, there are some studies in the literature where machine vision methods were used to detect the welding defects to help NDT experts:
[1] applied Discrete Wavelet Transform (DWT) on friction stir welding images and extracted the texture features. They classified the extracted features with Support Vector Machines (SVM) method and obtained 97% classification performance. In [2], surface defects were defined by image processing methods on friction stir welding images. Five different classes as voids, grooves, cracks, key-hole, and flash were used. [3] classified three different defect types as good welds, excess welds, and insufficient welds. During the classification process, the texture features were extracted by Gray-level Co-occurrence Matrix (GLCM) method. SVM method classified the extracted method. They used only 45 images and achieved 94.45% test performance. [4] detected the welding defects on Radiographic Testing (RT) images. They extracted the texture features on defects by Principal Component Analysis (PCA) method and classified the features with Linear Discriminant Analysis (LDA) technique. They classified 174 images into five different classes and achieved 91.12% classification performance. [5]

classified 1761 RT images into three classes as lack of penetration, incomplete fusion, and external undercut by k-Nearest Neighbor (kNN) and SVM methods. The best classification performance was obtained on lack of penetration defect as 98.03%.

This study has a dataset that consists of three classes, which have 2250 images in total. The texture feature extraction process was implemented by Grey-Level Co-occurrence Matrix (GLCM) [6], Discrete Wavelet Transform (DWT) [7], and Grey-Level Size Zone Matrix (GLSZM) [8] methods and different feature types were extracted. The extracted features were combined and a feature set that represents the image was created. The feature sets of all images were classified by ANN [9] technique. This paper has four parts. In the second part, the images and features are explained. The third part presents the classification results. The fourth part evaluates the results and discusses future studies.

2. Material and Method

The images in this study were taken on welding that was made by different synergistic welding machines in Hürsan Press Company. The images were taken under three different environments as the dark environment, daylight, and factory environment. Three types of defects, which were made by five different welders, were photographed in three different environments to increase the reliability of the proposed method. Because of five different welders, the dataset has data diversity. The images were taken with 3 mega-pixels snake camera, which is positioned with a right angle, from a 3cm distance. The images were obtained with 640x480 dimensions. The snake camera has six LEDs around itself to provide a high light on welding. NDT experts also use high white light to evaluate the type of defects. Figure 1 shows the welding in the factory environment and the apparatus, which was used to take photographs.

Figure 1. Factory environment and the apparatus used.

Figure 2 shows the sample images after image acquisition.

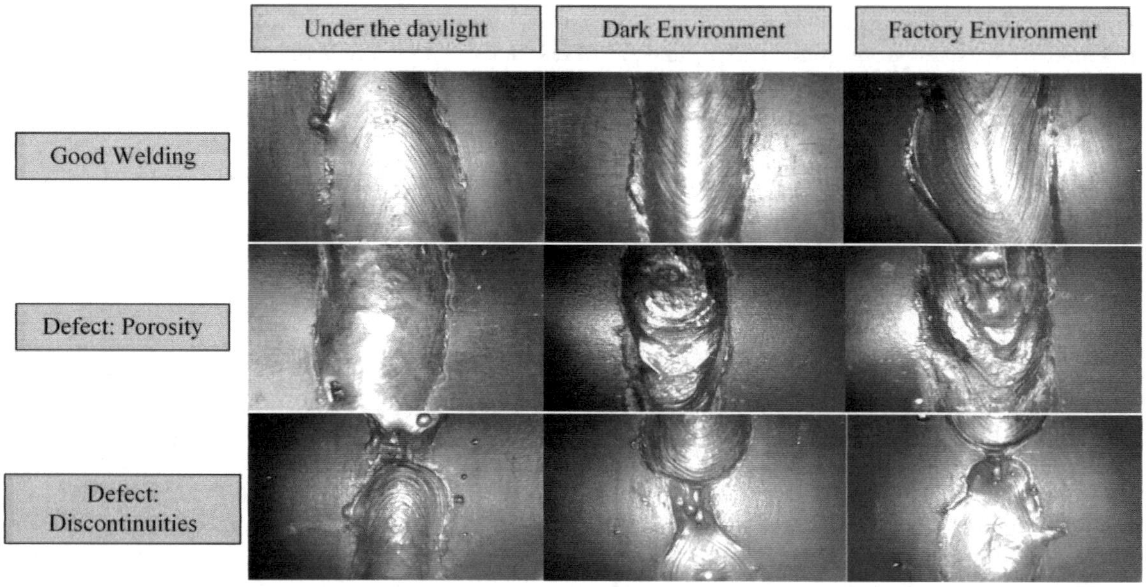

Figure 2. The images in the dataset that have different types of defects.

As shown in figure 2, all of the weldings were taken under different environments. All LEDs on the snake camera were used in full power in all environments during image acquisition. The reason is that NDT experts check the welding under high white light. Figure 2 shows the porosity defect, which is caused by insufficient protection and dirty workpieces. Discontinuities defects occur according to the ability of welder and whether the speed of wire riding is fast or slow. This study proposes a method to classify the welding defects automatically. In the proposed method, the defects are classified by artificial intelligence methods. The proposed method extracts the texture features on welding by GLCM, GLSZM, and DWT techniques. Then, the extracted features were used to train the ANN classifier. The scheme of the proposed method is presented in figure 3.

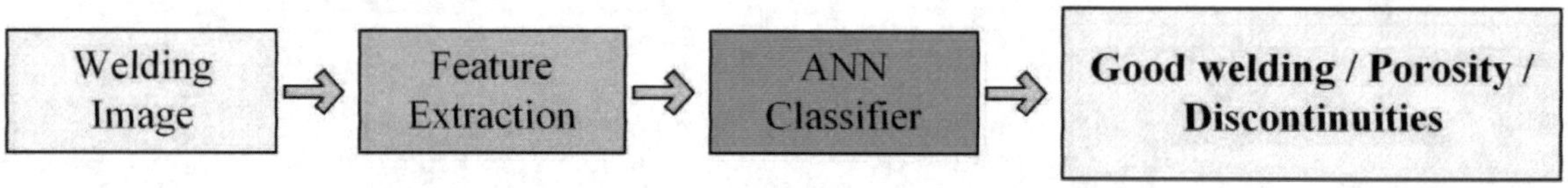

Figure 3. The scheme of the proposed method.

During the feature extraction process, 19, 13, and 24 features were extracted by GLCM, GLSZM, and DWT (db1 wavelet is used) methods, respectively. In the DWT method, the mean, standard variation, energy, entropy, skewness, and kurtosis features were obtained on four result images. All extracted features were combined and a feature set, which has 56 features, was obtained for one image. The system was trained with 56 features for each image.

3. Experimental Results

This section presents the classification results. During the classification stage, three different parameters in the ANN method were changed to evaluate the classification results. The learning rate (LR) parameter was changed between 0.1~1 (step by 0.1) and 2~10 (step by 1). The number of iteration (NoI) was changed between 5000~20000 (step by 1000), and the number of hidden neurons (NoHN) was taken as 6, 13, and 27. Three hidden layers were used in the neural network structure. Purelin and traingdx functions were used as activation and training functions, respectively. There could be an infinite number of experiments in the ANN structure; therefore, the experiments were implemented between the intervals given above. The neural network, which gave the best classification performance, was saved to use during the test stage.

The dataset has 750 images for each class. 500 and 250 images were used for training and test stage, respectively. During the training stage, the network was trained with 10-fold cross-validation method and tested by training and validation dataset, which were automatically created during the 10-fold cross-validation method. The network, which gives higher classification accuracy on validation dataset than training dataset, was saved. This provided to prevent the overfitting situation in the network. That means the network learned the dataset, did not memorize. The networks, which have higher validation performance, were used in the test stage and the obtained test performances were compared. The detailed scheme of the training and test stages are presented in figure 4. The best classification results and the parameters are presented in table 1.

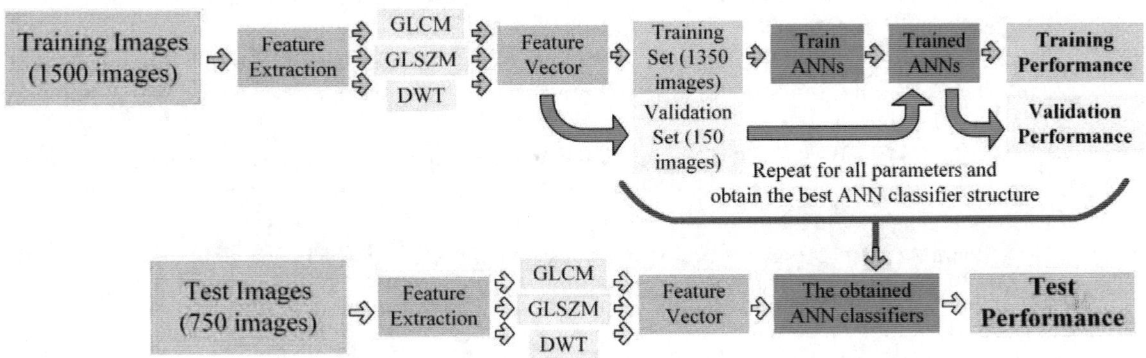

Figure 4. The detailed scheme of the training and test stages.

Table 1. The classification results for different ANN parameters.

Neural Network	**Data**	**Evaluation Metrics (%)**					**ANN Parameters**		
		SEN	**SPE**	**ACC**	**PRE**	**F-score**	**NoHN**	**NoI**	**LR**
Network 1	Training	90.47	95.11	93.48	91.08	90.60	56/**6**/3	11000	5
	Validation	90.60	95.16	93.56	91.16	90.69			
	Test	83.6	99.14	93.72	98.12	90.28			
Network 2	Training	91.24	95.54	94.03	91.82	91.36	56/**13**/3	16000	1
	Validation	91.47	95.67	94.19	92.17	91.57			
	Test	**85.2**	**99.15**	**94.31**	**98.16**	**91.22**			
Network 3	Training	88.53	94.11	92.11	89.44	88.71	56/**27**/3	9000	2
	Validation	88.87	94.31	92.36	90.04	89.04			
	Test	83.20	99.58	93.89	99.05	90.43			
Network 4	Training	90.21	95	93.30	90.91	90.35	56/**27**/3	20000	0.2
	Validation	90.27	95.03	93.35	91.05	90.40			
	Test	82.8	99.36	93.58	98.57	90			

Table 1 shows that the best test performance was obtained with Network 2 as 94.31%. Other networks also gave promising classification results in the proposed method.

4. Conclusions and Discussions

This study classified the welding images with high performance. The images of different weldings were taken in different environments to increase the reliability of the proposed method. The training duration of the ANN structure was obtained as 282 seconds, while the test duration was obtained as 0.98 seconds only. This shows that the proposed method detects the type of defect in shorter than one second.

As mentioned in the experimental results section, there could be done an infinite number of experiments to find the optimum classifier parameters. This study chose a specific interval for three different parameters and obtained the results.

IOP Conf. Series: Materials Science and Engineering **895** (2020) 012012 doi:10.1088/1757-899X/895/1/012012

Future studies will focus on classifying six different defect types. A smart system will be created to detect the defect with a high classification performance. The smart system will be implemented on hardware to use in Hürsan Press Company to help the NDT experts during the check process.

5. References

[1] Bhat NN, Kumari K, Dutta S, Pal SK, Pal S. Friction stir weld classification by applying wavelet analysis and support vector machine on weld surface images. Journal of Manufacturing Processes. 2015;20:274-81.

[2] Ranjan R, Khan AR, Parikh C, Jain R, Mahto RP, Pal S, et al. Classification and identification of surface defects in friction stir welding: An image processing approach. Journal of Manufacturing Processes. 2016;22:237-53.

[3] Shah HNM, Sulaiman M, Shukor AZ. Automatic Classification Vision Based System for Welds Joints Defects Using Support Vector Machine (SVMs). International Journal of Soft Computing. 2017;12(1):66-71.

[4] Mu W, Liu G, Liu P, Tian X. A novel classification approach of weld defects based on dual-parameters optimization of PCA and LDA. International Conference on Advances in Mechanical Engineering and Industrial Informatics; 2015: Atlantis Press.

[5] Moghaddam AA. Image processing techniques for classification of linear welding defects. 2015 2nd International Conference on Knowledge-Based Engineering and Innovation (KBEI); 2015: IEEE.

[6] Koyuncu H, Ceylan R, Asoglu S, Cebeci H, Koplay M. An extensive study for binary characterisation of adrenal tumours. Medical & biological engineering & computing. 2018:1-14.

[7] Shensa MJ. The discrete wavelet transform: wedding the a trous and Mallat algorithms. IEEE Transactions on signal processing. 1992;40(10):2464-82.

[8] Thibault G, Fertil B, Navarro C, Pereira S, Cau P, Levy N, et al. Shape and texture indexes application to cell nuclei classification. International Journal of Pattern Recognition and Artificial Intelligence. 2013;27(01):1357002.

[9] Haykin S. Neural networks: Prentice hall New York; 1994.

Acknowledgments

This study was funded by Hürsan Press Company and Konya KOSGEB branch.

IOP Conf. Series: Materials Science and Engineering **895** (2020) 012013 doi:10.1088/1757-899X/895/1/012013

Non-destructive Fertility Detection of Multiple Chicken Eggs Using Image Processing and Convolutional Neural Network

Wittaya Koodtalang[1] Thaksin Sangsuwan[1] and Apinai Rerkratn[2,*]

[1] Department of Instrumentation and Electronics Engineering, Faculty of Engineering, King Mongkut's University of Technology North Bangkok, Bangkok, Thailand.
[2] Faculty of Engineering, King Mongkut's Institute of Technology Ladkrabang, Bangkok, Thailand.

[*]Corresponding author: apinai@gmail.com

Abstract. This paper presents a non-destructive fertility detection of multiple chicken eggs in incubation industry based on image processing technique and convolutional neural network (CNN). The aim of this research were to design and implement the system for simultaneously distinguish multi eggs infertility from fertility one. LED light source setting up for illumining the 48 eggs, consisting of both egg types and randomly placed on a tray in dark box. In addition, a pre-trained CNN is performed to classify fertile and infertile eggs. All eggs is captured and processed to extract a region of interest (ROI) for each egg, generating large ROI egg images dataset which used to train and test the CNN model. A designed system is programmed using Python, operating on Windown7-64bit supported by OpenCV and Keras. Experimental results showed the accuracies for fertile incubated eggs detection between day 7 and day 9 reaches 100%. Meanwhile, eggs location has 100% of accuracy is also observed. Hence, the proposed technique are high reliability, high accuracy system and suitable to use in real application.

1. Introduction

Chicken production is the one of food industries which plays an important role to Thailand's purpose to be the kitchen of the world, one of crucial government policy. Thailand produced the chicken products over 2.1 million tons in 2016, placing the top five largest countries in the word by exporting the chicken meat [1]. However, the most important factor of chicken production is the number of day-old chicks that supply to all scale of this industry's operation. Incubation industry is the day-old chicks' supplier, a significant factor that highly influences the chicken industry. Incubation industry is totally dependent on the fertility and hatchability of eggs. Fertility is a ratio of fertile eggs to incubated eggs, while hatchability refers to a ratio of fertile eggs that hatch to all fertile eggs [2]. Early detection infertile and absence of development eggs cloud allow hatcheries to improve hatching rates, saving incubator spaces, handling cost and preventing the spread of infections from exploder eggs [3]. Candling hatching eggs during incubation are convenient tool for analyzing poor hatches with recognizing the dead embryos or clear eggs. In this method, the light source is illuminated and sharped to the eggs in dark box, resulting in the inner side of eggs can be visible and one can distinguish infertile eggs and early dead embryos from fertility one. If the inside of an egg is a black spot near the center alongside with some blood vessel, it is considered as a fertile egg, while the inside of an infertile egg is wholly clear without any blot [4]. Nowadays, the candling process is labor employing; the vast number of eggs are randomly candled about 5%. Nevertheless, the thousands of eggs per day

are checked by labors, leading to the loss of capability due to fatigue and sight mistake of employers. In addition, the separation between fertility and non-fertility of eggs must be operated by expert workers [5][6].
Recently, many researchers have been proposed the different methods for nondestructive automatic detection of fertile eggs. Although, the proposed methods using hyper spectral imaging [3][5][7], ultrasound [8], heart rate detection [9], oxygen flux detection [10], visible/near infrared transmittance spectroscopy [11] and thermal imaging [12], which are showed high accuracies and abilities to detect fertility in early, but those techniques are expensive and complicated. Low cost detecting fertility instruments using light-dependent resistor [13] and temperature sensor [14] has also been proposed. Like the methods that mentioned in above, the process for classifying between fertilized and non-fertilized of them are sophisticated. It can be seen that the automated systems using computer vision for detecting fertility have not only been obtained high accuracy, but also achieved low cost implementation [6][15][16]. However, only single egg can be inspected by the methods above. In spite of multi egg inspections simultaneously capability in ref [17], a low accuracy of 91.43% is obtained. Nowadays, the convolutional neural network is becoming more popular method to practically classify objects in image. Many researcher has been applied CNN in different fields, such as agriculture [18], medical [19], robotic [20], traffic enforcement [21] and so on. This paper presents multi fertile egg detection simultaneously based on image processing technique and CNN. The eggs location and sorting can be accomplished by image processing, producing resized images fed into pre-trained CNN model to classify between fertile and infertile eggs. CNN model is constructed by three convolutional layers and two fully connected layers. The following paper is organized as follows: overview of proposed model, eggs location and fertile eggs detection technique, experimental results and conclusion.

2. Overview of the proposed model

A block diagram of a proposed model demonstrates in Figure 1. Multi eggs are placed on a tray inside dark box. LED light source is illuminated and sharped to all eggs under a tray, enabling the inner side of eggs can be visible. It should be noted that 48 eggs are placed on a tray with 6 rows and 8 columns, resulting the proposed system can be inspected all of eggs simultaneously. A HD webcam is then captured eggs image with resolution of 1280x960 pixels. Captured image can be removed the noises with 5x5 Gaussian average filter and eroded pixels with 7x7 mask, called preprocessing task. The canny technique is applied to find the edges of each egg in the preprocessed image. Then, the locations of each egg are achieved by Hough circle transform. However, the proposed sorting algorithm is taken to rearrange each egg's location into correct order of rows and columns. It makes the detection system can be known in each location whether egg available or not. Each detected eggs area is cropped and resized corresponding to the size of input of pre-trained CNN model. Finally, the predictor classifies each egg images whether fertility or non-fertility, and label results related to eggs sorted location.

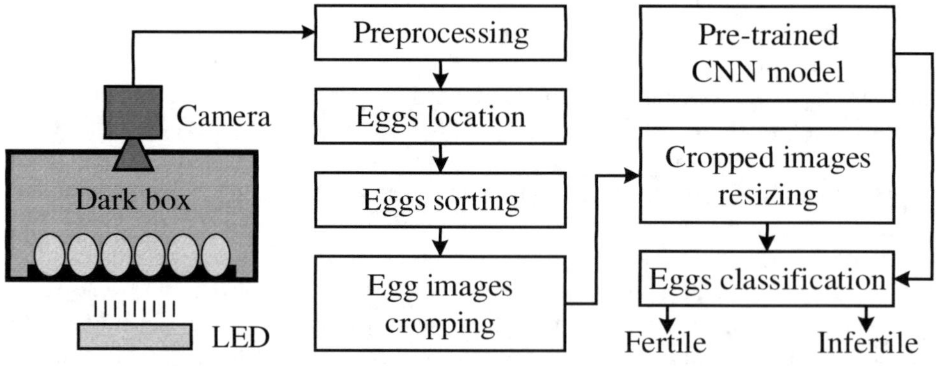

Figure 1. Block diagram of the proposed model.

3. Eggs location and fertile eggs detection

There are two main procedures of the proposed system for classifying between fertile and infertile eggs, each eggs location and fertile eggs detections. Image processing technique is applied to remove noise in acquired image and used Hough circle transform to locate each egg's position. The sorting algorithm is also used in this step. To detect fertile eggs, the state of the art machine learning called convolution neural network, the one method of deep learning (DL) is utilized, in order to predict the eggs whether fertility or non-fertility. The deep details for both processes will be explained in sequence.

3.1. Each eggs location

Figure.2 illustrates pseudo code of the proposed technique to locate each egg in captured image. In each loop, RGB image showed in Figure 3(a) is captured and converted into gray color to remove its dimensions. Both 5x5 Gaussian average filter and 7x7 erosion mask are used to reduce the noises in gray image. Then, the filtered image shows in Figure 3(b) can be detected the edges using canny technique, demonstrating result showed in Figure 3(c). All circles in edged image can be extracted by Hough circle transform, producing difference center positions and radiuses for each circle. To remove the detected circle that might be the noises, hence the detected circles must have radiuses greater than expected values. The filtered circles can be numbered by compute Euclidean distant between detected center and expected center, expressing as

$$DE_i = \sqrt{\left(cX_i - dX_i\right)^2 + \left(cY_i - dY_i\right)^2} \quad \text{when } i = 1,2,3,\dots,47,48 \tag{1}$$

where DE_i presents Euclidean distant of circle index i, (cX_i, cY_i) and (dX_i, dY_i) are the center position of expected and detected circle index i , respectively. It should be noted that the expected centers had defined before compute DE_i. Detected circle can be numbered with i number at which DE_i is minimized, indicating as

$$\underset{i}{\arg\min}\left(DE_i\right) = \left\{ i \mid DE_i = \min_i\left(DE_i\right) \right\} \quad \text{when } i = 1,2,3,\dots,47,48 \tag{2}$$

The result of circle detection and sorting is illustrated in Figure 3(d). Next, region of interest (ROI) for each egg are cropped corresponded to rectangles' position in previous step. Some example egg ROI image is demonstrated in Figure 3(e). Finally, each ROI images are resized related to the input size of CNN classifier, showed in Figure 3(f).

```
while true
  if exit then break end if
  Acquire RGB image
  Convert RGB image to gray
  Filtering gray with 5x5 Gaussian average filter
  Erode filtered image with 7x7 mask
  Detect edges in eroded image using canny
  Find all circles in edged image using Hough circle transform
  for cX, cY, radius in circles
    if radius > expected radius
      Sorting circle by comparing detected (cX,cY) and expected center
      Crop gray image as rectangle ROI
      Resize cropped image
      Feed resized image into classifier model
    end if
  end for
end while
```

Figure 2. Pseudo code of the proposed technique to positioned each egg.

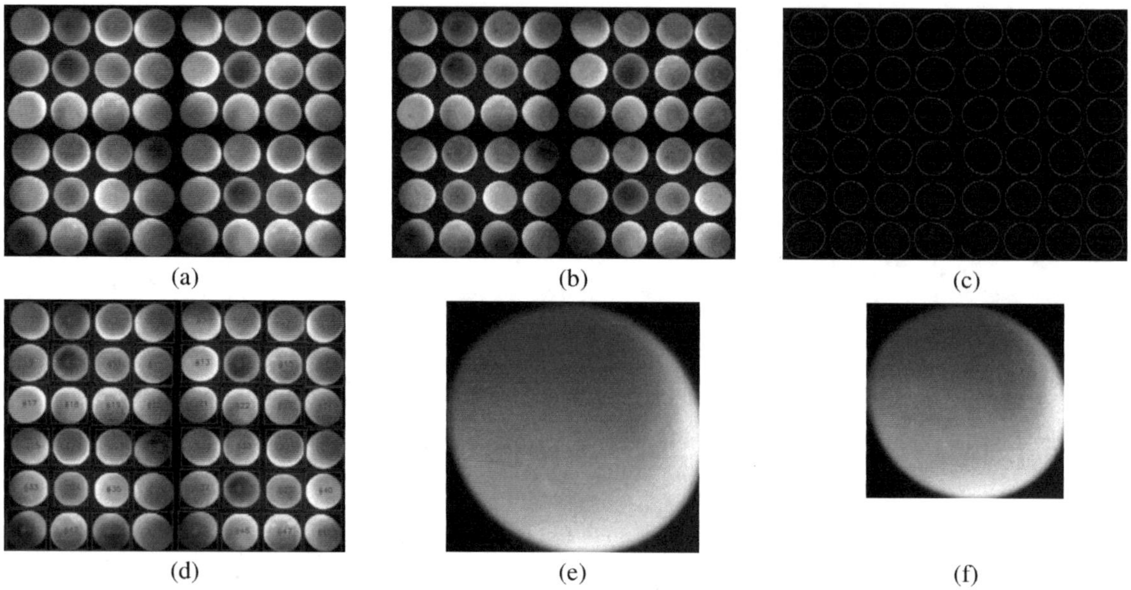

Figure 3. Each egg location procedure; (a) Acquired RGB image; (b) Filtered image; (c) Resulted image of edge detection; (d) Resulted image of each egg location and sorting; (e) Cropped image; (f) Resized image.

3.2. Fertile eggs detection

The classification between fertilized and non-fertilized of eggs uses CNN classifier model, constructing by three convolutional layers and two fully connected layers showed in Figure 4. It can be seen that three convolutional layers are supported by rectifier linear units (ReLU) and max-pooling like sandwich, it can be enhanced learning process. The input image constitutes segmented eggs of 64x64x3 pixel resolution. The first and second convolutional layers use 5x5 filters and they has 64 filter, while the third layer uses 3x3 filter with 128 filters. The third convolutional layer generates pooled output and feeds into the first fully connected layers. The dropout ratio of 0.5 is added to the inputs of the second fully connected layer, preventing over-fitting during training process. The output from dropout is then sends into Softmax classifier to classify the input image is whether fertile or infertile egg. Softmax activation function provides statistic that shows the probability of both categories. More details of training process will be described in section 3.3.

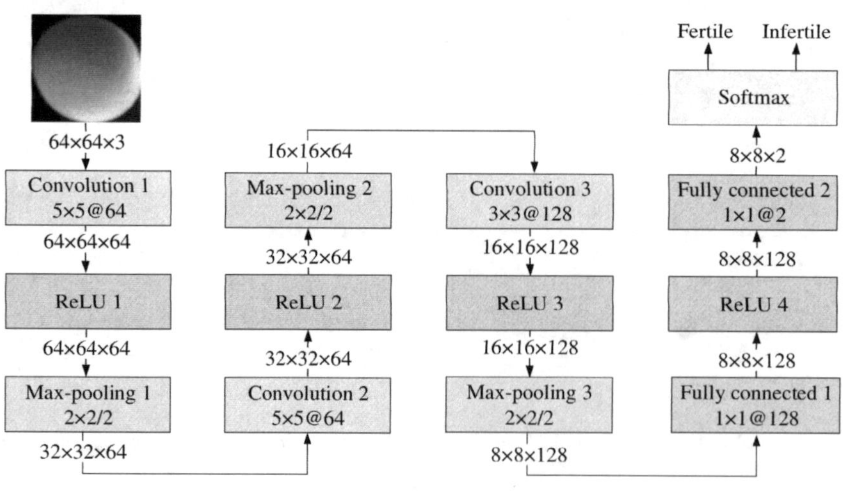

Figure 4. Architecture of the proposed classifier model.

3.3. Training the classifier model

The model for classifying the eggs is trained by optimizing the multinomial logistic regression objective using stochastic gradient descent (SGD). By initializing both learning rate and decay to be $1x10^{-1}$, while the number of epochs is defined to 30 and initialized the batch size equal to 32. The model is trained on Windown7-64bit with Intel CPU core i-7 2.0GHz processor, 8GB RAM, Python 3.7.3 and Keras 2.2.4 with Tensorflow 1.14.0 backend. The datasets to train the model consist of 313 and 603 images of fertile and infertile eggs, respectively. Meanwhile, the validation images have fertile eggs images of 75 and infertile eggs images of 95. Some fertile eggs images shows in Figure 5(a) and infertile eggs images are demonstrated in Figure 5(b). The best performances of both training and validation accuracies are 100%. It has training and validation loss of 0.1025 and 0.1147, respectively.

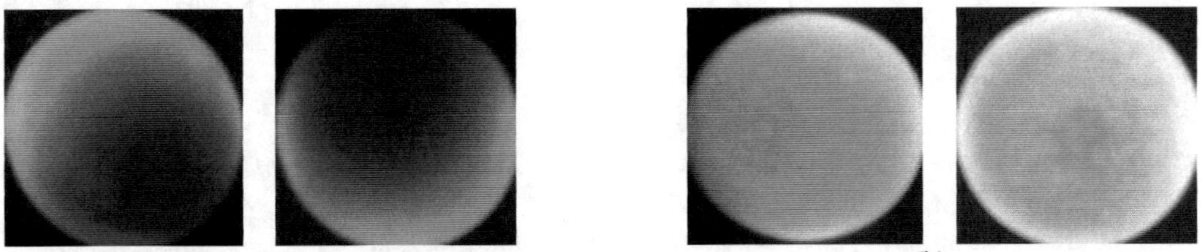

(a) (b)

Figure 5. Some images dataset; (a) Fertile eggs; (b) Infertile eggs.

4. Experimental results

To verify performances of the proposed model in Figure 1, experimental has been setup illustrated in Figure 6(a). The 240 eggs comprise 45 fertile (7-9 days) and 195 infertile eggs that the CNN pre-trained model never seen before, are used to verify the model. The 48 eggs mixed with both types of eggs which are placed randomly positions on a tray. The detection process use the sources as training process explained in section 3.3. However, OpenCV 4.1.0 is applied to find each egg positions in location process. The proposed model shows 100% accuracies of both location and fertile eggs classification for 240 samples, separating to 5 times classification. It should be noted that fertile eggs detection accuracy is compared with the result that achieved by expert of fertile eggs classification. Some detected result image is shown in Figure 6(b).

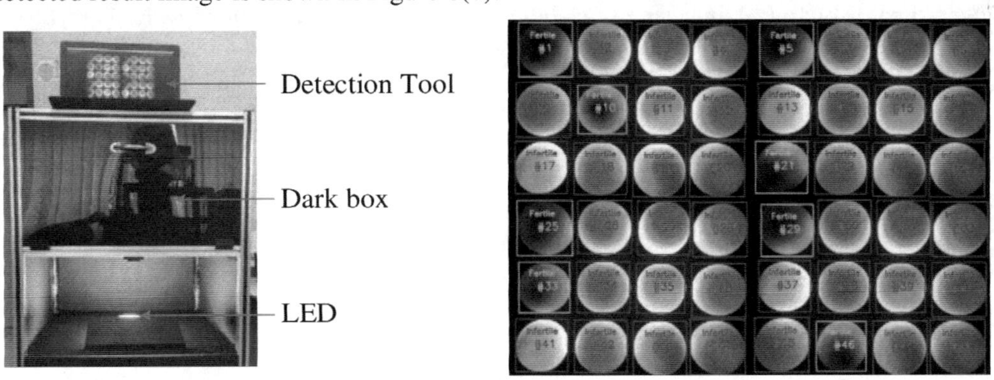

(a) (b)

Figure 6. Experimental; (a) Experimental setup; (b) Experimental result.

5. Conclusion

The classification between fertility and non-fertility of multi eggs is proposed in this paper. There are two main process, eggs location and fertile eggs detection. In location process, image processing technique is used to preprocessing image and cropped eggs ROI. The CNN pre-trained model is implemented for distinguish eggs non-fertility from fertility one. The experimental results show 100% accuracies of both location and fertile detection for 240 mixed sample eggs. In this paper, the fertile

IOP Conf. Series: Materials Science and Engineering **895** (2020) 012013 doi:10.1088/1757-899X/895/1/012013

eggs of day7-9 are inspected and compared the results with expert. Experimental results showed that the proposed technique can be suitably applied to incubation industry. However, early classification between fertile and infertile eggs of day 1- 6 will develop in the future.

6. References

[1] Chetchuda C 2018 *Thailand industry outlook* vol 20 (Bangkok: Krungsri research)

[2] King'ori A.M 2011 *Kenya. J.Poul.Sci.* **10** 483

[3] Lui L and Ngadi M.O 2012 *Canada. J.Food.Bio.Tech.* **6** 2053

[4] Ernst R.A, Bradley F.A, Abbott U.K and Craig R.M 2004 pub 8134 (California: University of California)

[5] Smith D.P, Mauldin J.M, Lawrence K.C, Park B and Heitschmidt G.W 2005 *Proc. 11th Euro. Symp. On Quality of Eggs and Egg Products (Doorworth: Netherlands)* p 176

[6] Mahdi H and Nacer F 2016 *Iran. J.Comp.Intl.Syst.* **9** 850

[7] Zhu Z, Liu T, Xie D, Wang Q and Ma M 2015 *Chaina. J.Agric & Biol.Eng.* **8** 69

[8] Eray O, Ilker H, Timur G and Banur B 2017 *Turkey. Indian.J.Anim.Res.* **51** 322

[9] Lei G, Yuzhou H, Zhitao X and Jiangtao X 2019 *Chaina. J.Appl.Sci.* **9** 1408

[10] Wang Q, Fu D, Ma M and Zhang T 2017 *Chaina. J.Agric & Biol.Eng.* **10** 243

[11] Jun D, Xiaoguang D, Yanlei L, Yankun P, Kuanglin C, Cuiying V and Xiuying T 2019 *(USAand Chaina). J.Coptr.Eltn.Argic.* **157** 471

[12] Chern-Sheng L, Po T.Y, Der-Chin C, Yih-Chih C and Chi-Hung L 2013 *Taiwan. J.Coptr.Eltn.Argic.* **91** 94

[13] Said E.A, Wael M.E and Ahmed A.E 2018 *Egypt. Misr.J.Ag.Eng.* **35** 199

[14] Tong Q, Romanini C.E.B, Exadaktylos V, McGonnell I.M, Berckmans D, Bahr C, Bergoug H, Roulston N, Guinebretière M, Eterradossi N, Verhelst R and Demmers T.G.M 2016 *(UK, Belgium and Frence).J.Livst.Sci.* **183** 19

[15] Dehrouyeh M.H, Omid D, Ahmadi H, Mohtasebi S.S and Jamzad M 2010 *Iran. J.Adv. Sci & Tech.* **16** 43

[16] Bhuvaneshwari M and Palanivelu L 2015 *India. J.Eng.Tech & Sci.* **2** 64

[17] Lean S.T, Emmanuel J.G.E and Ralph L.L 2018 *Proc. Int. TENCON (Jeju: Korea)* p 701

[18] Koodtalang W and Sangsuwan T 2019 *Proc. 1st Int. Symp. ICA-SYMP. (Bangkok: Thailand)* p 183

[19] Sivaramakrishnan R, Sameer K.A, Mahdieh P, Kamolrat S, Hossain M.A, Richard J.M, Stefan J and George R.T 2018 *(USA, Thailand, Bangladesh and UK). peerj.*458

[20] Jinwang W, Wei G, Ting P, Huai Y, Lin D and Wen Y 2018 *Proc. 21st Int. Conf. On Information Fusion (Cambridge)* p 439

[21] Yogameena B, Menaka K and Saravana P.S 2019 *India. J.IET.Intlg.Transp.Syst.* **13** 1190

IOP Conf. Series: Materials Science and Engineering **895** (2020) 012014 doi:10.1088/1757-899X/895/1/012014

Multi-objective Tourist Trip Design Problem in Chiang Mai City

Warisa Wisittipanich and Chanagan Boonya

Department of Industrial Engineering, Faculty of Engineering, Chiang Mai University, 239 Huay Keaw Rd., Suthep, Muang, Chiang Mai 50200, Thailand

warisa.o@gmail.com

Abstract. This study aims to solve a multi-objective tourist trip design problem (MO-TTDP) in order to simultaneously maximize tourist satisfaction and minimize total traveling cost. The goal is to identify optimal routing for tourist under time window and time horizon constraints. The mathematical model is first presented to find optimal routing under limited budget. However, the exact algorithm cannot find solution with multiple objectives simultaneously. Therefore, this study applies a metaheuristic, called Global Local and Near-Neighbour Particle Swarm Optimization (GLNPSO), to solve the MO-TTDP. Several local search strategies are also proposed to enhance the solution quality. The proposed algorithm is implemented on the real-case study in Chiang Mai City, Thailand. The experimental results show that the proposed algorithm yields a set of diverse and high quality non-dominated solutions.

1. Introduction

Tourist trip design problem (TTDP) is defined as a route-planning problem for tourists in order to create a trip composed of a set or sequence of points-of-interests (POIs). The main objective of TTDP is to select POIs (or tourist attractions) that best satisfy tourist preferences under a set of constraints (e.g., travelling budget, visiting time at each POI, time window of POIs, time horizon/day). Typically, the types of TTDP are classified based on the number of derived routings or the number of travelling days; 1) a single tour variants of TTDP and 2) multiple tour variants of TTDP. The single tour variants of TTDP aim at finding an optimal routing for one day trip and can be modelled as Orienteering Problems (OP). The multiple tour variants of TTDP determine multiple routings based on the number of tourist's visiting days and can be modelled as Team Orienteering Problems (TOP).

Damianos et.al [1] provided an intensive survey on solution approaches for solving TTDP with various sets of constraints. According to [1], TTDP is NP-hard in which exact algorithms can only find optimal solution for small-size problem with a small number of nodes. Thus, approximate algorithms such as heuristic and metaheuristic approaches are more preferable to deal with TTDP complexity especially when large number of nodes is considered. Tang and Miller-Hooks (2005) [2] proposed a Tabu Search (TS) heuristic with an adaptive memory procedure to improve the efficiency of the search for solving TTDP with a single tour. Archetti et al. [3] presented three metaheuristic solution techniques (TS and two variable neighborhood search, VNS) to solve TTDP. Ke et al. [4] proposed an Ant Colony Optimization-based algorithm (ACO) for TTDP. The results showed that the proposed ACO provided as good as the results obtained by [3] with faster computing time. Schilde et al. [5] introduced two metaheuristic approaches (ACO and an extension of VNS) for solving multi-objective TTDP. Later,

Bouly et al. [6] proposed another effective metaheuristic technique called a memetic algorithm (MA) with several local search techniques to solve the TTDP.
Muthuswamy et al. [7] differently tackled the TTDP by applying Discrete Particle Swarm Optimization (DPSO) to generate one day trip at a time. Some local search techniques were added to enhance the quality of solutions. The DPSO showed its superior performances to other seven heuristic algorithms. In [8], particle swarm optimization inspired algorithm (PSOiA) was proposed by Dang et al in 2013 to deal with TTDP. The results showed that the proposed PSOiA outperforms all previously proposed heuristics. Labadi et al. [9] developed a local search heuristic technique for the TTDP with time window based on variable neighborhood structure. Yu et al. [10] introduced a new variant of orienteering problem with time window in which several types of transportation within a single tour are taken into consideration. The result showed that PSO provides good solution quality with fast computing time.
This study presented an application of an effective metaheuristic, called Global Local and Near-Neighbour Particle Swarm Optimization (GLNPSO), to solve multi-objective TTDP (MO-TTDP). The objective is to design tourist optimal routes that simultaneously maximize tourist satisfaction and minimize total traveling cost. Several local search strategies are also proposed to enhance the solution quality. The remainder of this paper is organized as follows. Section 2 presents a mathematical model of the TTDP. Section 2 provides an application of GLNPSO algorithm to deal with the problem. The proposed algorithm is implemented on the real-case study in Chiang Mai City, which is one of the most popular tourist attractive cities in Thailand. Thus, the experimental results are reported in Section 4. Finally, the conclusion is given in Section 5.

2. Mathematical Model

The problem is mathematically formulated as a mixed-integer programming (MIP) as follows.

Index
i, j: Tourist attraction ($i, j = 2, 3, \ldots, N$) and accommodation ($i, j = 1$)
k : Travelling day ($k = 1,2,.., M$)

Decision variables

X_{ijk} : 1 if a tourist travel from attraction i to attraction j on day k or 0 otherwise
Y_{ik} : 1 if a tourist visits attraction i on day k or 0 otherwise
A_{ik} : Arrival time at attraction i on day k
PD_{ik} : Departure time from attraction i on day k
E_k : End time of tourist routing on day k

Parameter

N : Number of tourist attractions and accommodations; $n = \{1, 2\ldots, N\}$
M : Number of travelling days; $k = \{1, 2\ldots, M\}$
U_i, U_j : Constant value used to eliminate sub-tours
C_{ij} : Travelling cost from attraction i to attraction j on day k
RC_i : Travelling Cost at attraction i
S_i : Satisfaction score of attraction i
t_{ij} : Travelling time from attraction i to attraction j
R_i : Estimated time spent at attraction i
TW_{imin} : Minimum time window of attraction i
TW_{imax} : Maximum time window of attraction i
L : A large number
T_{max} : Maximum time horizon
T : Ready time
B : Budget

Objective function:

$$Max = \sum_i^N \sum_k^M S_i Y_{ik} \tag{1}$$

Constraints:

$$\sum_j^N \sum_K^M X_{ijk} \leq 1 \quad ,\forall_i ; i = \{2, \ldots, n\} \tag{2}$$
$$\sum_i^N \sum_K^M X_{ijk} \leq 1 \quad ,\forall_j ; j = \{2, \ldots, n\} \tag{3}$$
$$\sum_i^N X_{ijk} = Y_{jk} \quad ,\forall_j ; j = \{1,2, \ldots, n\} \tag{4}$$
$$\sum_j^N X_{ijk} = Y_{ik} \quad ,\forall_i ; i = \{1,2, \ldots, n\} \tag{5}$$
$$\sum_{j=2}^N X_{1jk} \leq 1 \quad ,\forall_k ; k = \{1,2, \ldots, m\} \tag{6}$$
$$\sum_{i=2}^N X_{i1k} \leq 1 \quad ,\forall_k ; k = \{1,2, \ldots, m\} \tag{7}$$
$$\sum_i^N X_{ipk} - \sum_j^N X_{pjk} = 0 \quad ,\forall_k ; k = \{1,2, \ldots, m\} \quad ,\forall_p ; p = \{1,2, \ldots, n\} \tag{8}$$
$$U_i - U_j + 1 \leq L \times \sum_k^M \left(1 - X_{ijk}\right), \forall_i \forall_j ; j \neq 1, i = \{1,2, \ldots, n\}, j = \{2, \ldots, n\} \tag{9}$$
$$\sum_i^N \sum_j^N \sum_K^M ((C_{ij} + CR_i) \cdot X_{ijk}) \leq B \tag{10}$$
$$\sum_i^N \sum_j^N (A_{ik} + R_i + t_{ij}) - L \times \left(1 - X_{ijk}\right) \leq A_{jk} \quad ,\forall_k ; k = \{1,2, \ldots, m\}, \forall_i \forall_j ; i,j = \{2, \ldots, n\} \tag{11}$$
$$A_{ik} \geq TW_{imin} \quad ,\forall_k ; k = \{1,2, \ldots, m\}, \forall_i ; i = \{1,2, \ldots, n\} \tag{12}$$
$$A_{ik} + R_i \leq TW_{imax} \quad ,\forall_k ; k = \{1,2, \ldots, m\}, \forall_i ; i = \{1,2, \ldots, n\} \tag{13}$$
$$DP_{ik} \geq A_{ik} + R_i \quad ,\forall_k ; k = \{1,2, \ldots, m\}, \forall_i ; i = \{1,2, \ldots, n\} \tag{14}$$
$$A_{1k} \geq T \quad ,\forall_k ; i = \{1,2, \ldots, m\} \tag{15}$$
$$\sum_i^N \sum_j^N (A_{ik} + R_i + t_{i1}) - L \times \left(1 - X_{i1k}\right) \leq E_k \quad ,\forall_k ; k = \{1,2, \ldots, m\} \quad ,\forall_i \forall_j ; i,j = \{2, \ldots, n\} \tag{16}$$
$$\sum_i^N \sum_j^N (A_{ik} + R_i + t_{ij}) - L \times \left(1 - X_{ijk}\right) \leq E_k \quad ,\forall_k ; k = \{1,2, \ldots, m\} \quad ,\forall_i \forall_j ; i,j = \{2, \ldots, n\} \tag{17}$$
$$E_k - A_{1k} \leq T_{max} \quad ,\forall_k ; k = \{1,2, \ldots, m\}, \tag{18}$$
$$x_{ijk}, y_{ik} \in \{0,1\} \quad ,\forall_k ; k = \{1,2, \ldots, m\}, \forall_i \forall_j ; i,j = \{1,2, \ldots, n\} \tag{19}$$

The objective of the model is to maximize tourist satisfaction from attraction scores as shown in equation (1). Constraints (2) and (3) state that there should be no more than one visit at tourist attraction *i* and *j*, respectively. Constraints (4) and (5) ensure the connection point between decision variable X_{ijk} and Y_{ik}. Constraints (6) and (7) ensure that the travelling trip on day *k* returns to the starting point. Constraint (8) ensures the continuity of the travelling trip on day *k*. Constraint (9) guarantee that the sub-tour is eliminated. Constraint (10) states that the travelling trip is arranged within the limit budget. Constraint (11) states that an arrival time at any attraction must be in time window. Constraints (12) and (13) guarantee that the visiting time at each attraction must be within time window. Constraint (14) states that the departure time from an attraction must be equal to or greater than its arrival time plus visiting time. Constraint (15) determine the starting time of a travelling trip on day *k*. Constraints (16) and (17) ensure that the departure time from the last visited attraction must be less than the end travelling time on day *k*. Constraint (18) ensures that the travelling time on day *k* is completed within maximum time horizon. Constraints (19) state that decision variables X_{ijk} and Y_{ik} are binary.

3. Application of GLNPSO with swap strategy

The original PSO algorithm was proposed by Kennedy and Eberhart in 1995 [11] in which its concept is originated from the schooling behavior of fish or flocking behavior of birds. The key concept of PSO is that each particle learns from the cognitive knowledge of its experiences (personal best, *pbest*) and the social knowledge of the swarm (global best, *gbest*) to guide the particle to better position. Several variants of PSO have been proposed to improve the performances of the original PSO. This study applies an extension of PSO, named GLNPSO, to solve the TTDP. Different from original PSO, GLNPSO, proposed by Pongchairerks and Kachitvichyanukul [12], incorporates multiple social learning terms

IOP Conf. Series: Materials Science and Engineering **895** (2020) 012014 doi:10.1088/1757-899X/895/1/012014

instead of using only single global knowledge. In particular, two additional social learning terms which are local best (*lbest*) and near neighbor best (*nbest*) are added in order to enlarge the search space and thus enhance the search efficiency. To increase exploitation capability, this study includes a local search technique called "swap" to intensify the search after some good solutions are found. The framework of GLNPSO with swap strategy is illustrated in Figure 1.

The framework of GLNPSO with swap strategy is developed to deal with multi-objective tourist trip design problem (MO-TTDP). The objective is to find tourist routing that simultaneously maximize tourist satisfaction and minimize total traveling cost. In this framework, solutions are expressed as a set of non-dominated solutions in which several routings are obtained under different budget. Consequently, decision alternatives are provided to tourists, and they can choose an option that best suits their goal.

For the encoding procedures, this study starts from defining the number of particle dimensions to be equal to number of attractions plus twice as much as the number of accommodation. Consider an example with one hotel and nine attractions. The dimensions of a particle are set to be 11 (9 + (2*1)). Each value in a vector dimension is initially generated with a uniform random number between 0 and 1. The dimensions of a vector are divided into two main parts; an accommodation part (dimension value is always equal to 0) and attraction part, as shown in Figure 2. To decode a particle dimension into a routing, a sorting list rule is applied to the attraction part to obtain a sequence of visiting. As shown in Figure 2, the travelling route is determined according to the order of sorted values in a particle dimension. The visiting at each attraction is examined against its time window and time horizon constraints. If the time window or time horizon constraint are violated, the next attraction will be considered. This process continues until a trip return to the accommodation.

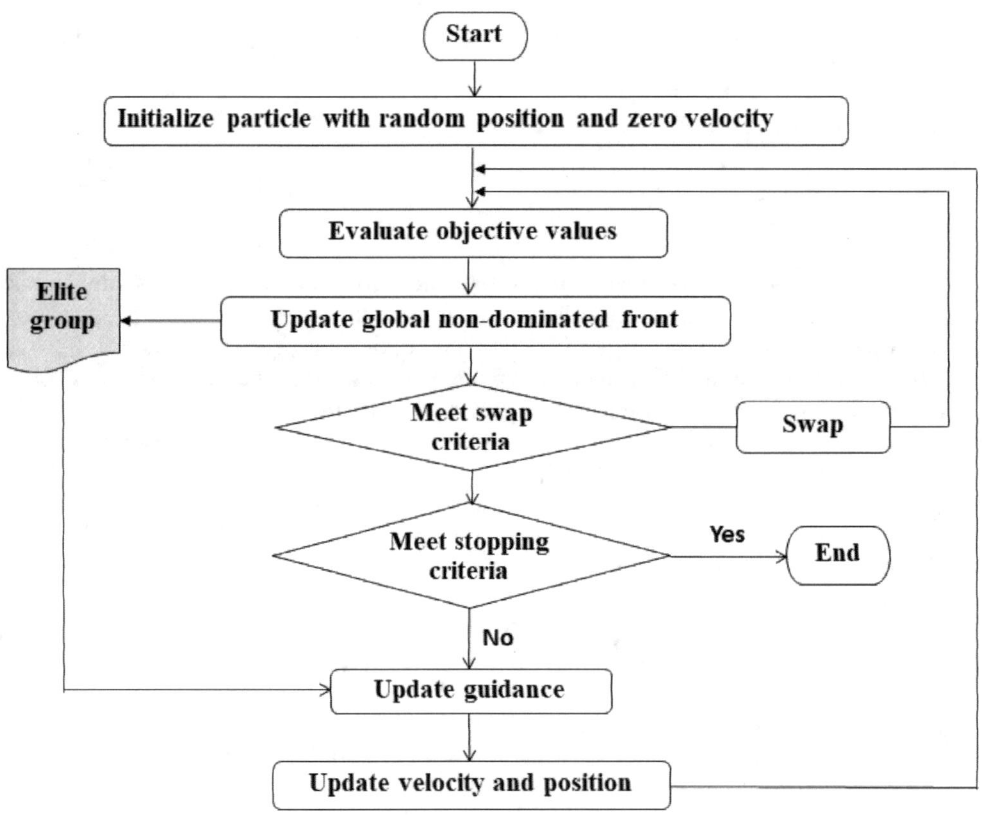

Figure 1. Framework of GLNPSO with swap strategy.

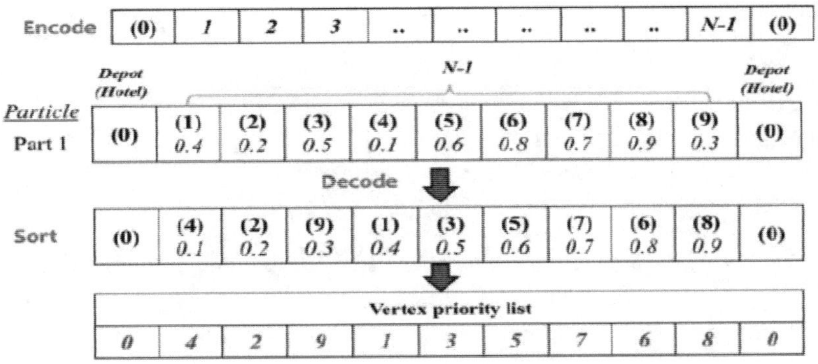

Figure 2. Solution representation.

This study implements swap technique to intensify the search once the swarm discovers some good solutions. The swap operation is applied when a specific number of iterations is met. Three swap strategies are proposed in this study and the explanation of each strategy are provided as follows.
Swap1: The idea of swap1 strategy is to exchange the position of attractions within one day trip. There are three possibilities that can occur which are 1) swapping between unvisited attractions, 2) swapping between visited attraction and unvisited attraction, and 3) swapping between visited attractions. The example of swap1 strategy is illustrated in Figure 3.

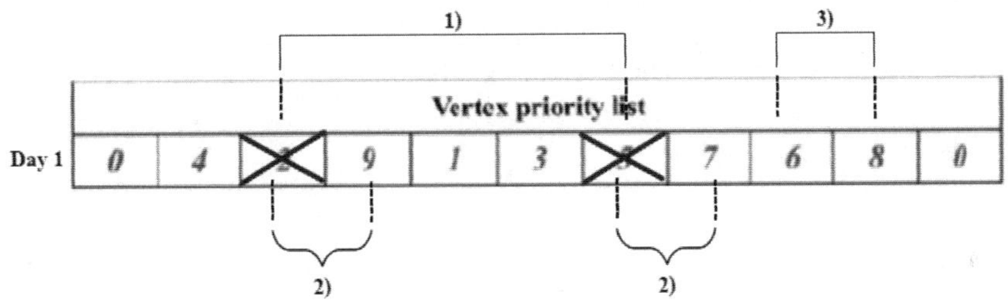

Figure 3. Swap1 strategy.

Swap2: Swap2 strategy is applied when a tourist plans to have two or more travelling days. This strategy exchanges the position of attractions from different travelling days. Similar to swap1, three possibilities; 1) swapping between unvisited attractions, 2) swapping between visited attraction, and unvisited attraction, and 3) swapping between visited attractions can occur in this case. The example of swap2 strategy is illustrated in Figure 4.

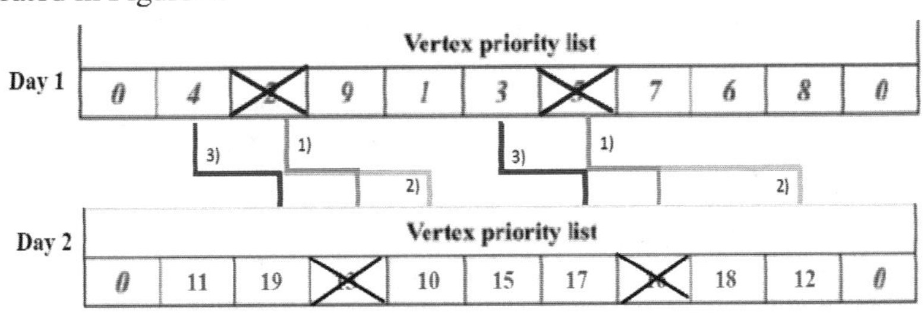

Figure 4. Swap2 strategy.

Swap3: Similar to swap2 strategy, swap3 is applied when a tourist plans to have two or more travelling days. However, swap3 combines the idea of swap1 and swap2 in which the attractions can be exchanged within the same travelling day and different travelling days. The example of swap3 strategy is illustrated in Figure 5.

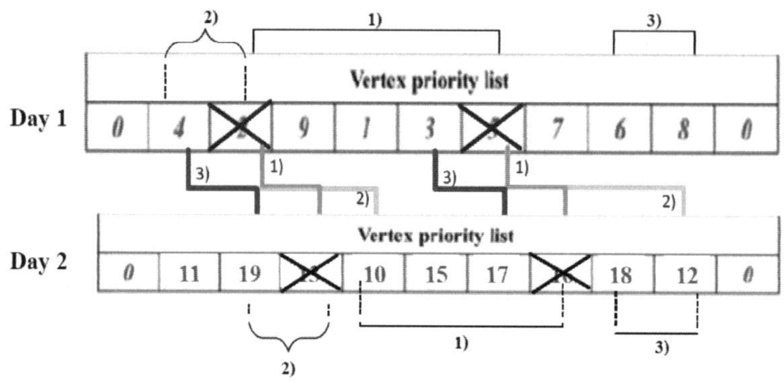

Figure 5. Swap3 strategy.

4. Experimental Results

4.1. Numerical Experiments

In this study, 10 problems are generated based on the real case study in Chiang Mai city, Thailand, for both single tour and multiple tours. The performances of GLNPSO with swap strategy are evaluated and compared with solutions obtained from LINGO optimization program and original GLNPSO. Table 1 show the comparison results when a tourist traveling by a local taxi which is one of the most popular transportation modes in Chiang Mai. It is noted that the objective values obtained LINGO are maximize tourist satisfaction.

Table 1. The result comparison between GLNPSO with swap strategy and LINGO

Instance	LINGO	GLNPSO no swap	GLNPSO with swap1	GLNPSO with swap2	GLNPSO with swap3
1	21	(21,120) (17,100)	(21,120) (17,100)	(21,120) (17,100)	(21,120) (17,100)
2	21.5	(21.5,420) (21,120) (17,100)	(21.5,420) (21,120) (17,100)	(21.5,420) (21,120) (17,100)	(21.5,420) (21,120) (17,100)
3	30	(30,1040) (29,600) (25,580) (24.5,280)	(30,1040) (29,600) (25,580) (24.5,280)	(30,1040) (29,600) (25,580) (24.5,280)	(30,1040) (29,600) (25,580) (24.5,280)
4	38	(38,2400) (24,2370) (29.5,1800)	(38,2400) (24,2370) (29.5,1800)	(38,2400) (24,2370) (29.5,1800)	(38,2400) (24,2370) (29.5,1800)
5	NA	(46.5,1270) (46,710) (45.5,550) (41.5,510) (41,420) (36.5,400)	(46.5,990) (46,730) (45.5,550) (41.5,430)	(46.5,1000) (46,700) (45.5,570) (41,350)	(46.5,1010) (46,730) (45.5,510) (41.5,430) (41,360)
6	NA	(46.5,2270) (46,700) (45.5,540) (45,450) (41,340)	(46.5,1000) (46,610) (45.5,540) (44.5,520) (41.5,430)	(46.5,1450) (46,780) (45.5,460) (45,450) (41,320)	(46.5,1080) (46,640) (45.5,530) (45,440) (40.5,380)
7	NA	(60,3770) (59.5,2720) (59,840) (54.5,820) (54,620)	(60,1510) (59.5,1340) (59,650)	(60,1530) (59,1250) (58.5,1110) (58,620)	(59.5,1050) (58.5,820) (54.5,800) (54,690)
8	NA	(77, 3840) (76.5, 1560) (76,1300) (72.5,1250)	(77,1590) (76.5,1410) (73,1370)	(77.5, 1820) (76.5,1300) (72.5,1230)	(77.5,5390) (77,2410) (76.5,1100)
9	NA	(93.5,6370) (93,5080) (90,4690) (89.5,3380) (89,2930) (88.5,1220) (87.5,1010)	(93,3560) (90,3550) (89.5,1070) (88.5,1000)	(93,3560) (90,3550) (89.5,1070) (8.5,1000)	(93,3560) (90,3550) (89.5,1070) (88.5,1000)
10	NA	(98,4940) (96.5,4380) (92,1690)	(97.5,3720) (97,3590) (89,2720)	(97.5,4540) (93.5,2220) (92,1510)	(97.5,2400) (94,1810) (93,1740)

It can be seen from Table 1 that all PSO algorithms are able to find optimal solutions of maximizing tourist satisfaction when compared with those from LINGO in all small- size problems (Instance 1-4). Additionally, the non-dominated solutions obtained from PSO offer a tourist decision alternatives with different travelling cost so that a tourist can select any solution that best suits their needs. When the problem size increases, while LINGO cannot find solution within a reasonable time, PSO algorithms are capable to provide a set of non-dominated solutions.

5. Conclusion

This research focuses on multi-objective tourist trip design problem (MO-TTDP). The goal is to design optimal tourist routing under time window and time horizon constraints with objectives to simultaneously maximize tourist satisfaction and minimize total traveling cost. The problem is first mathematically formulated as a mixed-integer programming model (MIP). Then, GLNPSO with swap strategy is applied since MIP cannot find optimal solution when the problem becomes more complex. The proposed algorithm is implemented on the real-case studies in Chiang Mai City, Thailand. The results show that the proposed algorithm is capable to yield good solution quality in all cases. Thus, this research can be used as an alternative approach to help tourists with better decision making in designing their trips under budget and time constraints. The future research includes extending the problem with more practical constraints or improving the algorithm performances by incorporating some strategies to better balance exploration and exploitation ability of the search.

6. References

[1] Damianos G, Charalampos K, Konstantinos M, Grammati P 2014 Survey on algorithmic approaches for solving tourist trip design problems *J. Heuristics*, **20** 3 pp 291-328

[2] Tang H and Miller-Hooks E 2005 A tabu search heuristic for the team orienteering problem *Comput Oper Res* **32** 6 pp 1379-1407

[3] Archetti C, Hertz A and Speranza M 2007 Metaheuristics for the team orienteering problem *J. Heuristics* **13** pp 49-76

[4] Ke L, Archetti C and Feng Z 2008 Ants can solve the team orienteering problem *Compt Ind Eng* **54** 3 pp 648-65

[5] Schilde M, Doerner K, Hartl R and Kiechle G 2009 Metaheuristics for the bi-objective orienteering problem *Swarm Intell* **3** pp 179-201

[6] Bouly H, Dang D-C and Moukrim A 2010 A memetic algorithm for the team orienteering problem *4OR Q J Oper* **8** pp 49-70

[7] Muthuswamy S and Lam S 2011 Discrete particle swarm optimization for the team orienteering problem *Memet Comput* **3** pp 287-303

[8] Dang D C, Guibadj R N, and Moukrim A 2013 An effective Pso-inspired algorithm for the team orienteering problem *Eur J Oper Res* **229** 2 pp 332-44

[9] Labadi N, Mansini R, Melechovsky J and Woler Calvo R 2012 The team orienteering problem with time windows: An LP-based granular variable neighborhood search *Eur J Oper Res* **220** 1 pp 15-27

[10] Yu V F, Jewpanya P, Ting C J and Perwira Redi A A N 2017 Two-level particle swarm optimization for the multi-modal team orienteering problem with time windows *Appl Soft Comput* **61** pp 1022-40

[11] Kennedy J and Eberhart R 1995 Particle swarm optimization *Proc. IEEE Inter. Conf. on Neural Networks (Perth)* pp 1942-48

[12] Pongchairerks P and Kachitvichyanukul V 2009 Particle swarm optimization algorithm with multiple social learning structures *Int. J. Oper. Res.* **6** 2 pp 176–94

Empirical study on influences of electromagnetic field on Hall Effect sensor for analysing oil lubricant deterioration

S Satthamsakul[1], W Sriratana[1] and L Sriratana[2]

[1]Faculty of Engineering, King Mongkut's Institute of Technology Ladkrabang, Bangkok, Thailand
[2]Faculty of Engineering, Ramkhamhaeng University, Bangkok, Thailand

E-mail: lerdlekha_s@ru.ac.th

Abstract. This study aims to present the procedure developed for detecting deterioration of oil lubricant under variation of electromagnetic field which was generated by a ferrite-core solenoid coil. The overall diameter of coil was 18.5 mm and the inductance was 22.421 mH from direct current stimulation. The results were then compared to those from frequency stimulation. The optimal efficiency of measurement systems was then assessed by real-time analyzing the metal particle contamination of 3 oil lubricants categorized by grade namely ISO 32, ISO 46 and ISO 68 as well as the viscosity under specific conditions which were flow rate and temperature ranging from 25 °C to 100 °C. The analysis was performed to compare the results from experiments of unused and used (approximately 6 months) oil lubricants. From experiments, the viscosity of oil lubricant was reduced when the temperature exceeded 50 °C. However, the stimulation of coil using 0.9 A direct current could clearly distinguish the difference between the unused and used oil lubricants with metal particle contamination comparing to other values of direct currents applied with 0.42 % of error. Also, the 0.5 kHz of frequency would be the most appropriate value for frequency stimulation with 0.38 % of error.

1. Introduction

Regular monitoring and maintenance of industrial machinery is necessary. Oil lubricant is one of the most important factors for mechanical performance to extend the lifetime of various parts and also increase the efficiency of the machine. Normally, the lifetime of oil lubricant is dependent on the type of oil and the hours of machine operation. However, in practice, oil lubricant could deteriorate sooner. There are many ways to check the properties for oil degeneration analysis such as ultrasonic measurement of viscosity of liquids [1], an IR-absorption sensor system for the determination of engine oil deterioration [2], influencing factors of viscosity measurement by rotational method [3], measurement and correlation of thermophysical properties of waste lubricant oil [4], photoacoustic measurement of liquid viscosity [5], oil contamination monitoring based on dielectric constant measurement [6], the viscosity measurement of oil lubricant based on Archimedes' principle including Newton's laws of motion using Hall Effect sensor [7] and application of magnetic field method for measuring lubricant viscosity [8].

This study used a Hall Effect sensor as an instrument measuring voltage variation due to metal particle contamination for analyzing oil lubricant deterioration. The intensity of electromagnetic field generated by coil which was stimulated by direct current and low frequency was also real-time assessed in order to inspect and analyze the properties of oil lubricant during operation effectively.

The results from analysis were then stored in the database for further assessment of machine conditions and maintenance plan.

2. Fundamental method

2.1. Principle of Hall Effect sensor
Galvanomagnetic effect is the phenomenon that electric charges can move in the medium under Lorentz force (F) of electromagnetic field on any electric charge (q) as illustrated in equation (1) [9]. When magnetic flux (B) acting in the direction perpendicular to the conductor sheet with the current flowing through, it can result in the deviation of the charge moving through the conductor that causes the potential different voltage or Hall voltage (V_H) between the two poles perpendicular to the direction of the current flow as shown in equation (2). [10]

$$F = qE + q[v \times B] \tag{1}$$

$$V_H = \frac{R_H}{d_1} IB \tag{2}$$

where E is the Hall field, v is the velocity of electron due to electric field, R_H is the Hall coefficient, I is the current through each coil and d_1 is the thickness of Hall Effect sensor

2.2. Viscosity and fluid dynamic
Viscosity (η) is an important parameter. It directly relates to the oil lubrication efficiency especially for industrial machinery that can contribute to high machine performance and can operate at full capacity. Equation (3) shows the relation of viscosity and time period when metal particles start moving through the solenoid coil until reaching saturation point under variation of the temperature due machine operation in real time. [11]

$$\eta = t_{moving} \frac{F'}{A} \rho_T \tag{3}$$

where t_{moving} is the time period when the particles start moving through the solenoid coil until reaching saturation point (sec), ρ_T is the density of fluid flow on south direction (km/cm^2), F' is the force caused by flow (N) and A is the area of the pipe where the fluid flows through (cm^2).

3. Structure of the testing system

3.1. Measurement processing of oil lubricant
The system designed for analysis of oil lubricant deterioration can be divided into 2 parts: 1. the part which can real-time simulate the flow of oil lubricant while the machine is working, consisting of oil pipeline (∅15 mm), motor pump, flow control valve, flow meter, heater (1.5 kW), PID controller for temperature and thermocouple type K with measurement, and 2. the measuring and processing part including solenoid coil, Hall Effect sensor, signal conditioning and computer for processing and storing data as shown in Figure 1.

3.2. Sensing part and placement of Hall Effect sensor
In this study, the quality of oil was evaluated by creating an electromagnetic field of the solenoid coil at the specified time (ON = 60 sec and OFF = 60 sec) to determine the speed of metal particles accumulating in the cross-section of the ferrite core until reaching the saturation point. The amount of metal particles caused by corrosion and abrasion of industrial machine parts during the operation was directly proportional to the voltage at the output of the Hall Effect sensor. The highest electromagnetic field intensity was at the Hall generator located between the cross-sectional area of the ferrite core and oil pipeline as shown in Figure 2. The output voltage from measurement was then used to analyze the amount of metal particles contaminated in the oil lubricant as well as the viscosity of the oil lubricant.

The results could indicate the quality of the oil lubricant being used and also estimate the service life of the oil lubricant in order to reduce the risk of damage that may occur to industrial machinery.

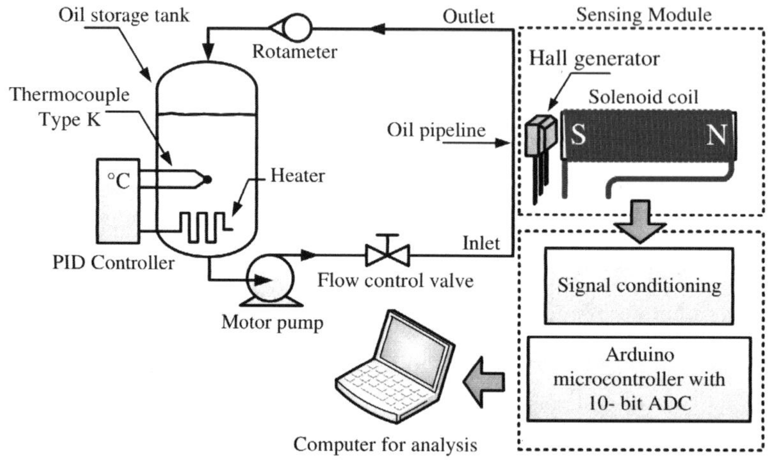

Figure 1. Block diagram of oil lubricant deterioration system.

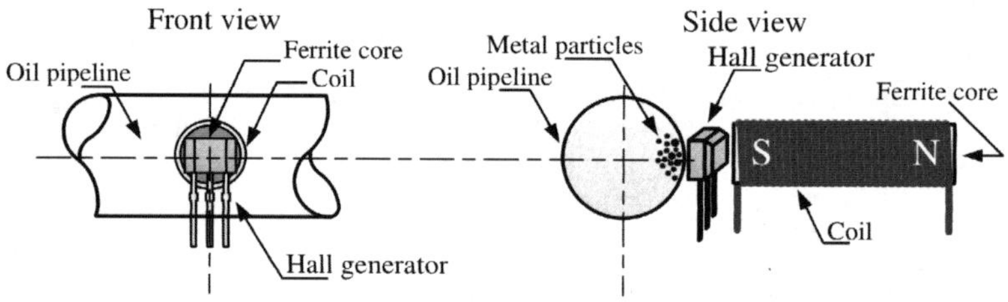

Figure 2. Installation of Hall Effect sensor located between the solenoid coil and the oil pipeline.

3.3. Experimental conditions

This study used hydraulic oils No. ISO 32, ISO 46 and ISO 68 as a case study to distinguish between the unused oils and the 6-months-used oils by generating electromagnetic fields using different stimulation methods under specific conditions. The sample of testing oils is as illustrated in Figure 3. The testing oil was placed in the tank with a heater to vary the oil temperature from 25 °C to 100 °C (increasing by the step of 10 °C) similarly to the real machine. Oil was pumped into the 15 mm diameter oil pipeline to measure the flow rate as well as the viscosity. The Hall generator positioned between the solenoid coil and the oil pipeline was used to measure and compare the amount of metal particles contaminating in the unused and used oils.

Figure 3. Sample of hydraulic oil No. ISO 32.

4. Electromagnetic field generation

The electromagnetic field generation using the solenoid coil excitation (cylindrical shape, diameter = 9.75 mm, and length = 56 mm), copper wire (SWG 22, 712 turns) with an inductance of 22.421 mH to enhance the Hall Effect sensor performance was calibrated with the measurement of reference metal particles (average diameter = 216 µm) and analyzed 5 times by a scanning electron microscope (SEM). The unused oil without contamination of metal particles were used as the reference voltage (V_{ref}) to be compared with the unused oils with metal particles in the amount of 0.2 g, 0.3 g, 0.4 g and 0.5 g, respectively. The relationship between the output voltage of Hall Effect sensor and the amount of metal particles is as shown in Figure 4. From the picture, an ISO 32 oil lubricant was tested at 32 °C with a magnetic field stimulated by a direct current of 0.9 A. From experiment, it was observed that the output voltage of the Hall Effect sensor would increase directly in proportion to the amount of metal particles and rise significantly to the saturation point when the amount of metal particles increased.

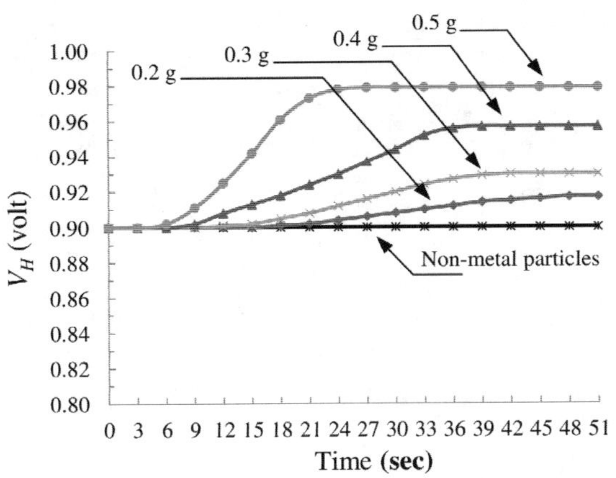

Figure 4. The relationship between the output voltage of the Hall Effect sensor and the reference amount of metal particles.

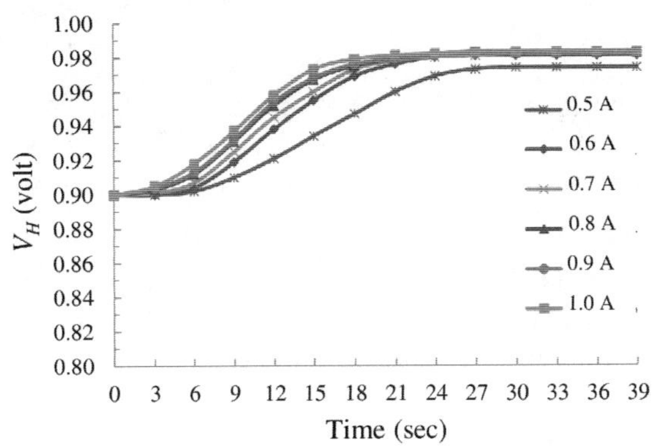

Figure 5. Comparison of different direct current stimulations with a sample of ISO 32 hydraulic oil at operating temperature of 50 °C.

4.1. Experimental results from direct current stimulation

The generation of the electromagnetic field by direct current stimulation was varied from the input current of 0.5 A to 1 A consisting of 0.5 A, 0.6 A, 0.7 A, 0.8 A, 0.9 A, and 1 A (increasing 0.1 A interval) under specific temperature of 50 °C. From experiment, it was found that the electrical current suitable for excitation of the solenoid coil was 0.9 A due to a high concentration of electromagnetic field generated at the cross-section of the ferrite core. Also, this electromagnetic field concentration was adequate for Hall generator to be operated without heat accumulating at the coil comparing to that excited by electric current of 1 A. Figure 5 is a sample test of an ISO 32 hydraulic oil used at operating temperature of 50 °C with an average estimating from 6 times of repeatability. The testing results of a sample of ISO 32 hydraulic oil used under temperature variation from 25 °C to 100 °C when flowing through the heater. Considering the flow rate of oil within the pipeline flowing through Hall Effect sensor, it can be observed that the viscosity of used oils could be reduced similarly to that of the hydraulic oil number ISO 46 and ISO 68 as shown in Figure 6.

4.2. Experimental results from frequency stimulation

The generation of electromagnetic fields using frequency stimulation was to assess the appropriate frequency for experiment by comparing the results of three frequencies: 0.5 kHz, 1 kHz and 1.5 kHz at 30 °C. The optimal frequency for solenoid coil excitation with the highest intensity electromagnetic field generator at the cross-section of the Ferrite core was at low frequency of 0.5 kHz. Comparing to direct current stimulation, it can be noted that frequency excitation results in a higher voltage output from the Hall Effect sensor than that of direct current stimulation. As a result, distinguishing metal particles contaminated in oil lubricant using frequency excitation can be well performed and contributed to the better accurate results of oil lubricant deterioration analysis. Figure 7 illustrates a sample test of the hydraulic used oil number ISO 32 with an average estimating from 6 times of repeatability. The testing results of ISO 32 hydraulic used oil under temperature variation from 25 °C to 100 °C. Considering the flow rate of oil through the Hall generator placed in The position between the solenoid coil and the pipeline, it can be observed that the viscosity of used oil would be reduced similarly to that of the hydraulic oils No. ISO 46 and ISO 68 that could directly affect the efficiency of the lubricating oil. Details are as shown in Figure 8.

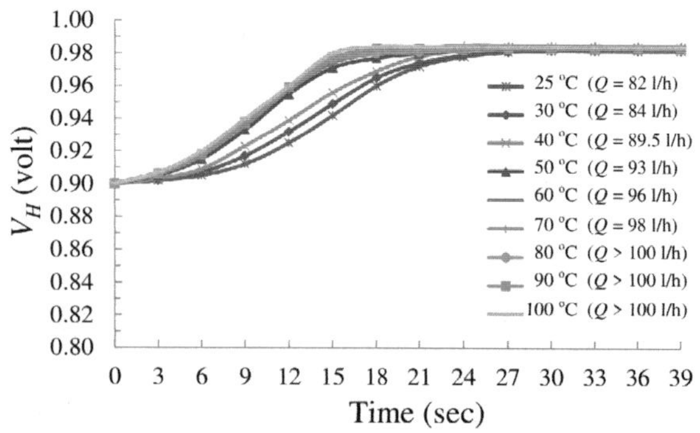

Figure 6. Test results of ISO 32 hydraulic oil used at temperature ranging from 25 °C to 100 °C with direct current stimulation of 0.9 A.

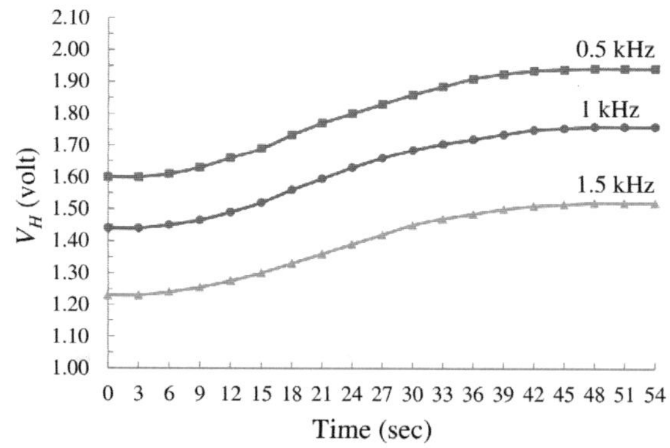

Figure 7. Comparison of the results from different frequency excitations of the hydraulic used oil No. ISO 32 at 30 °C.

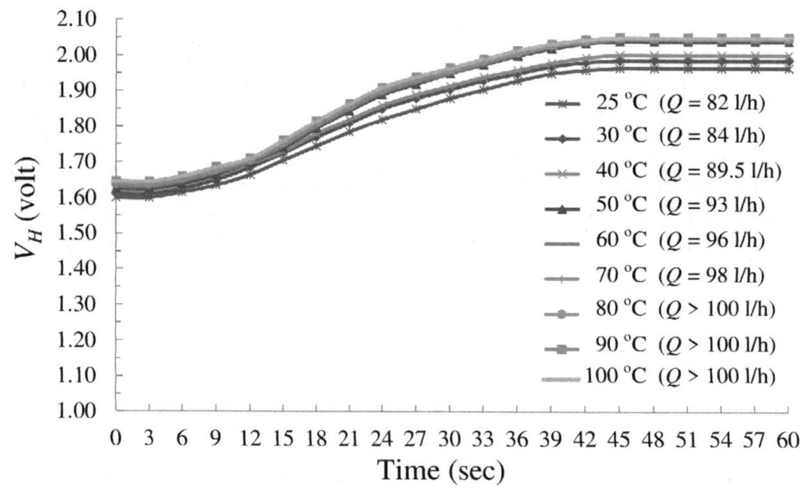

Figure 8. Testing results of the hydraulic oil No. ISO 32 used at temperature ranging from 25 °C to 100 °C with frequency stimulation of 0.5 kHz.

5. Conclusion and discussion

This study aims to compare the efficiency of the electromagnetic field generated by direct current stimulation and frequency stimulation in order to enhance the optimal efficiency of the Hall generator as the output voltage from the Hall Effect sensor is directly proportional to the amount of magnetic field intensity. Considering the excitation of the coil to create an electromagnetic field, the low frequency stimulation can increase the efficiency of the Hall Effect sensor due to higher Voltage Hall by more than 20 % without heat accumulation in the solenoid coil compared to the direct current stimulation. The frequency stimulation can also reduce the heat of the solenoid coil by about 48 % on average. However, when compared to other types of engine oil, such as fully synthetic SN 0W-20 of a 2,000 cc car with a 9-years lifespan, use of approximately 80,000 km, and engine oil changing period of 10,000 km, it can be observed that the viscosity of fully synthetic engine oil would increase proportionately with the temperature. It is because the engine oil has the ability to wash away soot from combustion caused by the engine's sword that results in increase of the viscosity. For the lubricant used for hydraulic machinery, the viscosity would decrease when the temperature rises due to oxidation reaction. Therefore, the analysis of oil lubricant or engine oil deterioration by considering the viscosity should also consider the type and nature of work of machinery or engine. For instant, the deterioration of the oil used for lubricating machinery without internal combustion systems would result in reduction of the viscosity. On the other hand, the deterioration of the oil used for lubricating engines with internal combustion systems that the oil can lubricate and wash away the soot from the combustion would contribute to increase of the viscosity.

6. References

[1] Buiochi F, Higuti R T, Furukawa C M and Adamowski J C 2000 Ultrasonic Measurement of Viscosity of Liquids *IEEE Ultrasonics. Symposium,* vol 1 (Puerto Rico: USA) pp 525-528

[2] Agoston A, Otsch C, Zhuravleva J and Jakoby B 2004 An IR-Absorption Sensor System for the Determination of Engine Oil Deterioration *Proceedings of IEEE, Sensor* vol 1, pp 463-466

[3] Zhang Y, Huang M, Kan Y, Liu L, Dai X, Zheng G and Zhang Z 2018 Influencing Factors of Viscosity Measurement by Rotational Method *Polymer Testing 70* pp 144-150

[4] Pinheiro C T, Pais R F, Ferreira A G M, Quina M J and Gando-Ferreira L M 2018 Measurement and Correlation of Thermophysical Properties of Waste Lubricant Oil *The Journal of Chemical Thermodynamics* vol 116 pp 137-146

[5] Lou C and Xing D 2010 Photoacoustic Measurement of Liquid Viscosity *Applied Physics Letters* vol 96

[6] Dingxin Y, Xiaofei Z, Zheng H and Yongmin Y 2009 Oil Contamination Monitoring Based on Dielectric Constant Measurement *International Conference on Measuring Technology and Mechatronics Automation* vol 3 pp 249-252

[7] Sriratana W, Nakklinkoon S and Satthamsakul S 2018 Test of Oil Lubricant Viscosity by Using Hall Effect Sensor *The 4th International conference on Engineering, Applied Sciences and Technology (ICEAST 2018)*, (Phuket: Thailand)

[8] Sriratana W and Murayama R 2013 Application of Magnetic Field Method for Measuring Lubricant Viscosity *Materials Science Forum* vol 750 pp 116-124

[9] Popovic R 1989 Hall-effect devices *Sensor and Actuators* vol 17 pp 39-53

[10] Rangan C S, Sarma G R and Mani V S V 2004 Instrumentation: Devices and System *Tata McGraw-Hill* (Publishing Company Limited: New Delhi)

[11] Sriratana W and Murayama R 2013 Measurement of the Lubricant Properties Using Hall Effect Sensor: A Study on Contamination and Viscosity *Engineer* vol 4 No. 5 pp 386-393

Acknowledgments

Authors wish to thank Faculty of Engineering, King Mongkut's Institute of Technology Ladkrabang and Faculty of Engineering, Ramkhamhaeng University Thailand for supporting this research.

IOP Conf. Series: Materials Science and Engineering **895** (2020) 012016 doi:10.1088/1757-899X/895/1/012016

Design and simulation of a novel MEMS based microfluidic circulating tumor cell (CTC) detection system for a lab on a chip device

M P Suriyage[1,*], P V K Asanka[1], T M G C S P Cooray[1], D L F M Liyanage[1], R A N I Pushpakumara[1], D K Hendavitharana[1], W W A T I Fernando[1], W H P Sampath[2] and Y W R Amarasinghe[1,2]

[1]Department of Mechanical Engineering, Faculty of Engineering, University of Moratuwa, Katubedda 10400, Sri Lanka
[2]Centre for Advanced Mechatronic Systems, University of Moratuwa, Katubedda 10400, Sri Lanka

[*]mpsuriyage@gmail.com

Abstract. This paper proposes a novel microfluidic system for a lab on a chip device which includes three major systems: a microchannel, micromixer and a droplet generator. The novel system proposes electroosmotic fluid flow control combining droplet generation and immunocapturing based CTC separation. ANSYS Fluent is used to optimize the fluid flow parameters, droplet size and justify the mixing capability of the micromixer. COMSOL Multiphysics simulations justify the integration of the electroosmotic fluid flow control for precise generation of droplets and optimization of dimensional parameters of microchannels, followed by a fabrication method for the microfluidic system.

1. Introduction

According to the 'National Vital Statistics Reports' for 2016 and 2017, cancer has identified as the second highest cause of death in the USA [1],[2]. Cancer cells are described as cells that divide relentlessly which in turn forms solid tumors and flood the blood with abnormal cells. Therefore, early stage cancer detection is a vital necessity to reduce deaths caused by cancer. Generally, cancer diagnosis is done through the extraction of tissue from the affected area (solid biopsy). Tissue extraction through biopsy is invasive, time consuming and expensive. Furthermore, cancer state monitoring is important, hence to give a better insight into the evolution of tumor dynamics during treatment and disease progression. A liquid biopsy is the other option which is the analysis of biomarkers in a non-solid biological tissue. There are several biomarkers that have been identified. CTCs, circulating tumor DNA (ctDNA) and exomes are three of the most common markers [3]. CTCs are the most commonly captured biomarker to detect cancer as it provides a unique identification of the type of cancer and the progression stage as it contains distinct morphological and phenotypic features. The concentration of CTC in the patient's blood can be measured in counts. The number of cells is countable compared to the exosomes.

There are few techniques to capture CTCs from whole blood. A positive selection of anti-bodies using anti-EpCAM was selected as the technique to follow in this paper [4]. Based on the available traditional methods, Lab on a Chip (LoC) devices have been developed to perform the separation process on a single chip rather than using complex integrated sub systems. LoC devices do not require bulky setups and those systems can provide quick test results and able to manipulate a single cell for further diagnostics and reliable observations.

Microfluidic CTC separation is done in two different methods: physical property based and immunocapturing based. Physical property-based cell separation techniques are inexpensive and does not require additional surface markers. But physical properties like size and density can create overlap regions with other blood cells at extreme limits and cause false results [4]. Physical based separation requires complex geometries which are difficult to fabricate. Immunocapturing is the other method that is used to separate CTCs in microfluidic systems. Immunocapturing facilitates downstream analysis of cancer cells such as proliferation, genotyping and drug sensitivity tests. Most immunoaffinity methods make use of the Epithelial Cell Adhesion Molecule (EpCAM) protein for immune interactions.

CTC capturing chips utilizing both separation methods are available in literature. 'CTC-chip' captures viable CTCs with high purity in a single step from non-pre-treated blood samples [5]. 'HB-chip' [6] isolated CTC clusters successfully using transverse flows within the micro channel to realize maximum collisions between cancer cells and the antibody-coated chip surface. 'CTC-iChip' is designed to obtain deterministic lateral displacement (DLD) followed by immunomagnetic separation of bead-labelled white blood cells (WBCs) and unlabelled CTCs [7].

Presented solution contributes towards the development of non-invasive methods for cancer risk prediction. Finding the number of CTCs available in a certain blood volume is difficult but the exact count is critical to get the maximum reliability of results. The aim of this research is to obtain efficient isolation of CTCs through droplet formation. This leads into identifying the threat of cancer and saves time by eliminating the requirement for performing various other tests which delays the consequential treatments. Proposed microfluidic system design can also be modified to capture different biological cells for several other diagnostic applications.

This paper discusses the microfluidic system design for immunocapturing based CTC detection. Section 2 presents the detail design of the microchannel, micromixer and the droplet generator. Section 3 proposes a fabrication method for the microfluidic system. Section 4 presents the results which are optimized through simulations.

2. Proposed MEMS based microfluidic system for CTC detection

The microfluidic solution proposed in this paper contains three reservoirs for blood, CD326 anti-EpCAM microbeads and oil as illustrated in figure 1. HFE-7500 fluorocarbon oil is selected for the droplet generation since it is biocompatible [8]. All three fluids are driven as electroosmotic flows. Blood and the microbeads are mixed using a micromixer which contains magnetic particles that are attracted by the CTCs. The blood and magnetic bead mixture are formed into droplets such that a droplet may contain only one CTCs. Then a magnetic system is used to separate the droplets which contains CTCs. These cells can be further observed through fluorescent microscopic techniques [9].

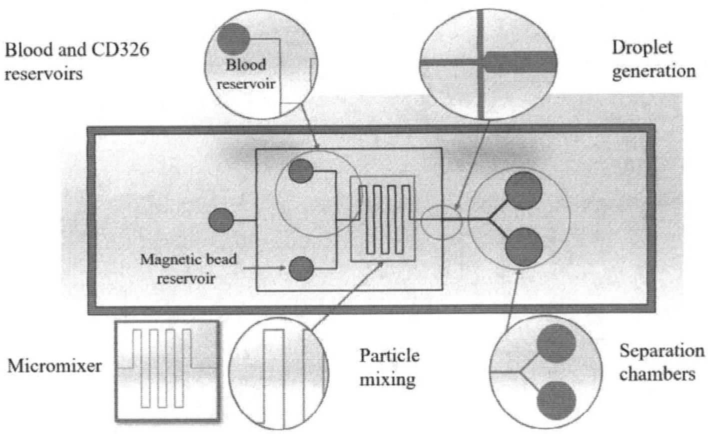

Figure 1. Designed microfluidic system.

IOP Conf. Series: Materials Science and Engineering **895** (2020) 012016 doi:10.1088/1757-899X/895/1/012016

2.1. Microchannel design

Microchannels are fabricated with different cross sections [10]. The most common types are rectangular, square, circular, half circular, U-shaped and gaussian beam shaped microchannels. The microfluidic system shown in figure 1 is designed using a square shape micro channel with a T-junction to transport blood and anti-EpCAM microbeads to the micromixer. Two separate square cross section paths are designed for the oil flow to generate droplets.

Microfluidic platforms have capillary driven, electroosmotic, electrowetting and pressure driven fluid flow systems. Electroosmotic fluid flows are generated naturally when a potential is applied and can be controlled using voltage and flat symmetric flow. Electroosmotic flow is much more controllable compared to other common techniques [11]. Therefore, the designed micro droplet generator uses electroosmotic methods for fluid flow control. Electroosmotic flow simulations performed in COMSOL Multiphysics were based on the required velocities for the droplet generation and mixing of fluids.

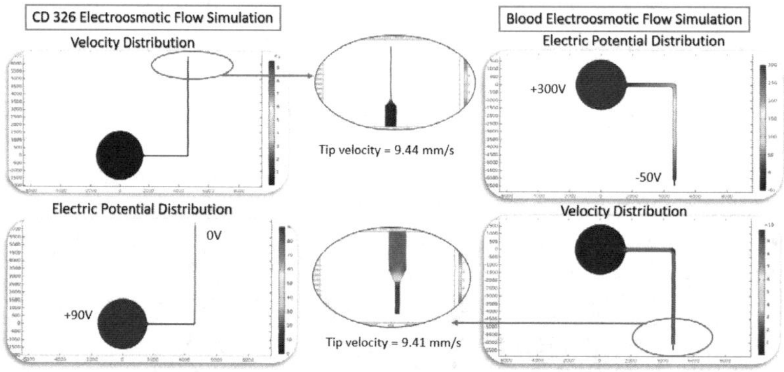

Figure 2. Electroosmotic flow simulations.

Desired velocities within the microfluidic channel is obtained by optimizing the voltages at the reservoir and junctions of the microfluidic system. Figure 2 shows the obtained tip velocities for corresponding potential drops along the microchannel. These velocities satisfy the velocity requirements for mixing and droplet generation of the complete microfluidic system.

2.2. Micromixer design

Micromixers are classified as passive and active micromixers [12]. Depending on fabrication feasibility and cost, a passive mixer is used for the initial design. With the simulation results Considering the mixing efficiency, a square wave shaped convergence-divergence based micro mixture is designed [13]. Convergence-divergence points increase the efficiency of the mixing process. For an accurate magnetic based separation, a 20:1 mixing ratio between blood and CD326 anti-EpCAM microbeads is recommended [14]. Designed micromixer enhances molecular diffusion and chaotic performance for mixing purposes. Depending on the required velocities from the droplet generation, mixing process was simulated using ANSYS Fluent. According to figure 3 convergence divergence geometry, length and cross-sectional dimensions are optimized to have a mixing ratio closer to the recommended.

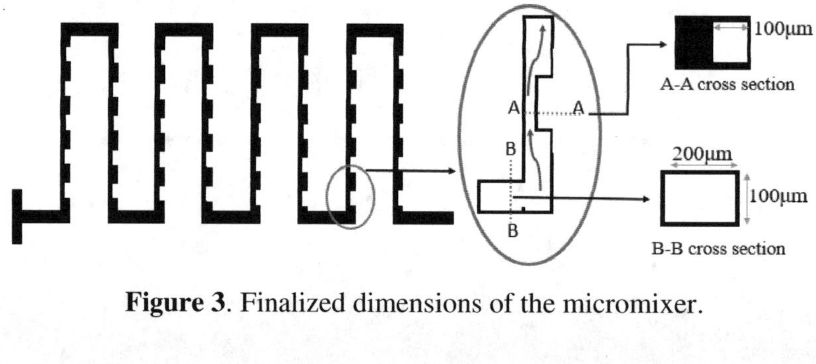

Figure 3. Finalized dimensions of the micromixer.

2.3. Droplet generation design

There are several active and passive techniques to generate fluid droplets [15]. In passive techniques such as squeezing, dripping, jetting and tip-streaming as shown in figure 4, the dispersed fluid is introduced into the continuous fluid to generate droplets. Active techniques require additional energy input through external sources, and it provides another level of controllability for droplet formation. Active techniques are categorized further as the additional force driven and intrinsic force driven.

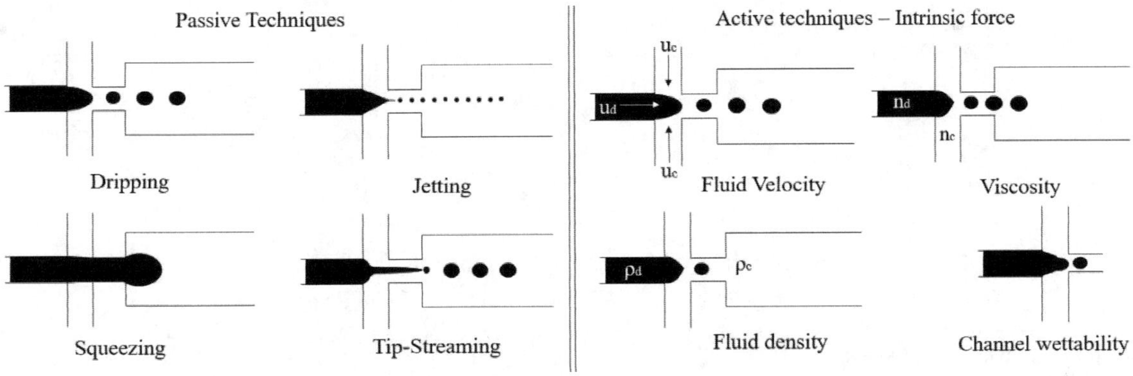

Figure 4. Passive and active techniques of droplet generation.

Generated droplets in the system contains blood and magnetic particles. CTCs have a diameter around 20 μm [4]. Droplet diameter should be less than 40 μm to ensure that it does not contain two CTCs and should be greater than 20 μm to contain one CTC. Therefore, the droplet size is decided to be around 30 μm.

Simulations are carried out using blood and HFE-7500 fluorocarbon to select the most suitable technique for the blood droplet generation. Three different geometrical designs are compared. Fluid velocity based active technique and the double T-section geometry are selected to generate desired size droplets. Geometry 2 generates droplets closer to the desired size whereas geometry 1 and geometry 3 results in undesirable droplet formations as shown in figure 5.

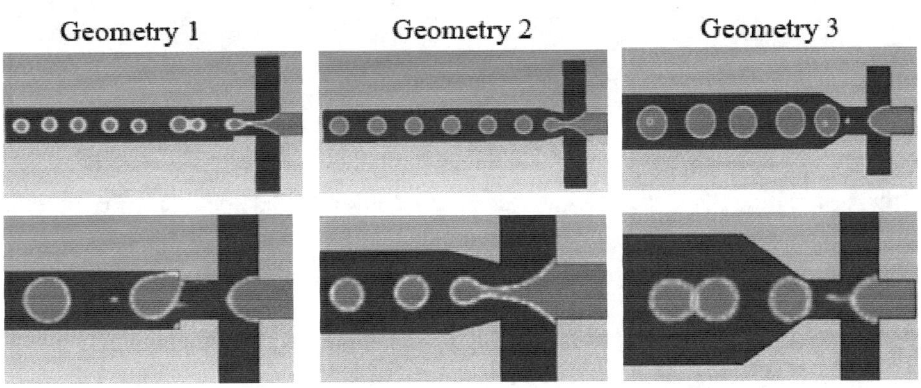

Figure 5. Comparison of different geometries for droplet formation.

3. Fabrication method

The entire microchip mould is designed on a silicon substrate according to the steps shown in figure 6. The first negative photoresist of SU-8 2050 [16] is spin-coated on silicon with a 50 μm thickness and patterned using photolithography. After photolithography, a second SU-8 negative photoresist layer is deposited for 2 mm in thickness and patterned to create the boundaries of chip batch production [17]. The PDMS layer is deposited with Si elastomer curing agent with a 10:1 ratio. PDMS is then peeled off from the mould and finally bonds with a glass of 2 mm thickness through glass-oxygen plasma bonding [18].

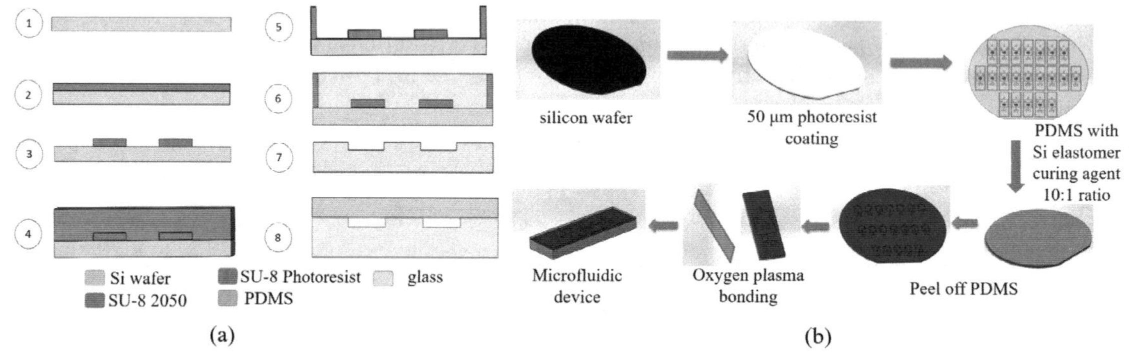

Figure 6. Schematic of microfabrication on silicon substrate (a) device fabrication steps (b) microfluidic system batch fabrication steps.

4. Results and Discussion

Microchannel dimensional parameters were optimized to obtain the optimum velocities, mixing ratio and droplet diameter. Finalized values for the microchannels are included in table 1. Micromixer design is optimized by using a convergent divergent point which minimizes the cross-section area of the mixing micro channel by half. This improves the mixing of blood and CD326 anti-EpCAM microbeads to a degree where the mixing ratio is 15:1 at the mixer output. Total length of the micro mixer is 70mm.

Table 1. Finalized dimensional and velocity values of the microchannel.

Fluid section	Width of cross section (µm)	Width of cross section end (µm)	Length of microchannel (mm)	Velocities of inlets to mixer and droplet generator (mm/s)	Voltages between the reservoir and inlet (V)
CD 326	50	2.5	9.5	9.44	90 - 0
Blood	200	50.0	9.5	9.41	300 – (-50)
Oil	50	50.0	46.5	20.10	350 - 0

Droplet diameter is determined by observing the blood volume fraction of two deviated points with respect to the centre line. The diameter of the droplet is observed to obtain the optimum velocity combination of blood and oil.

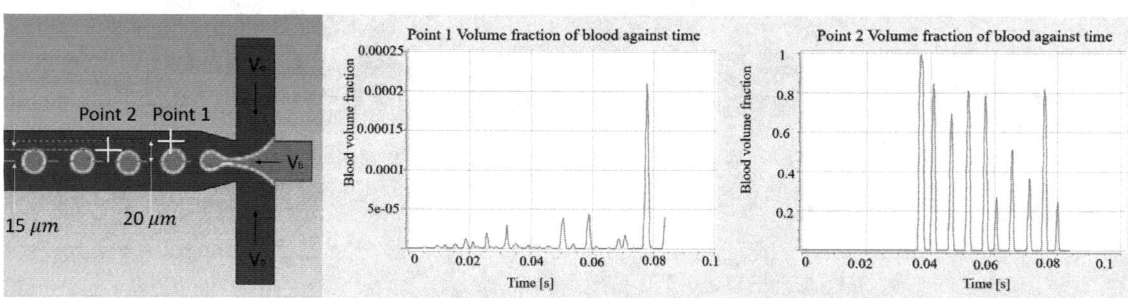

Figure 7. Droplet diameter observation.

The volume fractions at point 1 and point 2 which are 15µm and 20µm apart from the centre line, are obtained to determine the droplet diameter. For the optimum velocity combination of blood and oil: V_b

and V_o, the volume fraction of point 1 is less than 0.00025 and point 2 is greater than 0.8 as shown in figure 7. Hence, the droplet diameter is deduced to be greater than 30 μm and less than 40 μm. Optimum flow velocities of blood (V_b) and oil (V_o) were obtained as 10 mms^{-1} and 20 mms^{-1} to generate droplets of diameter around 30 μm. Therefore, the blood mixture and oil velocities of 10 mms^{-1} and 20 mms^{-1} were used to perform all other connected simulations.

5. Conclusion

ANSYS Fluent simulations supported the optimization of fluid flow parameters to get desired size droplets and the analysis to justify the mixing capability of the designed micromixer. COMSOL Multiphysics simulations aided the design process to optimize critical dimensional parameters of the microfluidic channels and justified the integration of electroosmotic fluid control to the system to obtain fluid flow velocity requirements. It would be interesting if future research could be conducted on the design of a magnetic system which is used for separation of droplets with CTCs. The additional integration of a microscopic system to count the number of CTCs available in blood would complete the microfluidic system for cancer risk prediction.

6. References

[1] Heron M 2018 Deaths: Leading Causes for 2016 *National Vital Statistics Reports* vol 65 no 5
[2] Heron M 2019 Deaths: Leading Causes for 2017 *National Vital Statistics Reports* vol 68 no 6
[3] Marrugo-Ramírez J, Mir M and Samitier J 2018 Blood based cancer biomarkers in liquid biopsy: A promising non-invasive alternative to tissue biopsy *International Journal of Molecular Science* **19** 2877
[4] Khajvand T Liu W Shi Y Song Y Tian T Wang S Yang C Zhu Z Zou Y 2017 Enrichment and single-cell analysis of circulating tumor cells *Chem. Sci* **8** pp. 1736–1751
[5] Nagrath S *et al* 2007 *Isolation of rare circulating tumour cells in cancer patients by microchip technology* vol 450 pp 1235-1239
[6] Sheng W, Ogunwobi O O, Chen T, Zhang J, George T J, Liu C and Fan Z H 2013 Capture, release and culture of circulating tumor cells from pancreatic cancer patients using an enhanced mixing chip *Lab on a Chip*
[7] Ozkumur E *et al* 2013 Inertial Focusing for Tumor Antigen-Dependent and-Independent Sorting of Rare Circulating Tumor Cells *Science Translation Medicine* **5** 179
[8] Ali M Digman M A Kang D K Zhang K 2014 Rapid detection of single bacteria in unprocessed blood using Integrated Comprehensive Droplet Digital Detection *Nat. Commun* **5** p. 5427
[9] Chen P Frenkel E Hoshino K Huang Y Y Lane N Liu H Sokolov K Wu C H Zhang J X L 2015 Screening and Molecular Analysis of Single Circulating Tumor Cells Using Micromagnet Array *Sci. Rep* **5** pp 16047
[10] Kumar S and Prakash S 2015 Fabrication of microchannels: a review *Journal of Engineering Manufacture* **229** 8 pp 1273-88
[11] Yang C Zhang W Zhao C 2017 Dynamic Electroosmotic Flows of Power-Law Fluids in Rectangular Microchannels *Micromachines* **8** p 34
[12] Cai G, Xue L, Zhang H and Lin J 2017 A review on micromixers *Micromachines* **8** 274
[13] Chen X, Li T and Li X, 2016 Numerical research on shape optimization of microchannels of passive micromixers *IEEE Sens. J.* **16** 17 pp 6527–32
[14] StraightFrom ^{TM} *Whole Blood CD326 (EpCAM) MicroBeads Human*
[15] Zhu P and Wang L 2017 Passive and active droplet generation with microfluidics: a review *Lab on a Chip* **17** pp 34-75
[16] Mata A, Fleischman A J and Roy S 2006 Fabrication of multi-layer SU-8 microstructures *J. Micromech. and Microeng.* **16** pp 276-84
[17] Zhang J 2002 LIGA mold insert fabrication using SU-8 photoresist *LSU Master's Theses*
[18] Jha S K, Joo G, Ra G, Lee H H and Kim Y 2011 Development of PCR microchip for early cancer risk prediction *IEEE Sens. J.* **11** 9 pp 2065-70

Acknowledgments
The Authors would like to express their gratitude to the Centre for Advanced Mechatronic Systems, University of Moratuwa for their advices and comments for the success of the research. Furthermore, authors would like to thank the academic staff and the research assistants of Mechatronics system Engineering Stream, University of Moratuwa. Special thanks for University of Moratuwa for their financial assistance.

AUTHOR INDEX